Before adding down, look for easy combinations.

8 7 2 → 10 +3 → +10 —— —— 20	3 6 3 → ☐ +4 → +☐ —— ☐	4 6 7 → ☐ +3 → +☐ —— ☐

4 4 5 +2	7 4 4 +3	4 2 6 +8	5 9 2 +1
9 8 2 +1	7 1 3 +9	4 1 3 +5	5 6 5 +4

Find the sums.

3	9	1	6	4
1	9	8	7	2
8	1	9	2	4
+2	+5	+3	+4	+7
7	8	4	2	5
8	4	3	1	6
6	1	7	7	0
+1	+3	+6	+7	+4
8	5	1	1	3
7	4	5	0	5
5	2	8	6	5
+2	+6	+6	+1	+8
2	6	0	5	7
5	2	6	3	9
4	3	3	9	4
+8	+9	+9	+7	+9

Add down, then up to check your answers.

1	1
2	2
3	3
+4	+4
10	10

7	7
4	4
9	9
+1	+1

6	6
5	5
1	1
+2	+2

3	3
7	7
1	1
+8	+8

9	9
6	6
8	8
+3	+3

4	4
8	8
3	3
+2	+2

7	7
5	5
6	6
+4	+4

5	5
9	9
4	4
+6	+6

4	4
7	7
2	2
+3	+3

Addition

35 24 +10	44 12 +41	13 22 +12	23 42 +21
30 34 +32	52 15 +22	21 12 +31	12 25 +31

24 22 +33	23 43 +20	31 10 +42	22 53 +20
25 10 +44	23 24 +22	32 15 +21	10 15 +20

Adding large numbers.

6135 +3602	1824 +4152	4534 +3463
4041 +4354	5207 +1722	2134 +6450
3232 +4733	3048 +5641	1535 +7214
1260 +6738	3523 + 6154	3240 + 2706
3143 +5320	2401 +5406	2400 +2106

Solve these problems.

1. On a walk in the meadow I saw ...
   9 crows
   3 meadowlarks
   6 pheasants
   How many birds did I see? ________

2. Water birds like to land on our pond. Last week I saw 11 ducks and eight geese. How many birds did I see? ________

3. My sister raises pigeons. She has eight now. How many will she have if I give her six more for her birthday? ________

4. Mom has chickens in the backyard. The hens lay eggs everyday. This morning she gathered nine white eggs and seven brown eggs to put in a special cake she is making. How many eggs did she use? ________

5. If you are very still at night, you can hear owls as they go hunting. Monday I saw 3 owls, Friday I saw 8 owls, and last night I saw 4 owls. How many is that in all? ________

6. I saw a nature program on television. It showed the colorful birds living in the rain forest. 8 brightly colored parrots were sitting in one tall tree. Then 9 more came flying up and perched on the branches. How many parrots did I see in the tall tree? ________

The more we dry, the wetter we get.

93 – t	90 – s	26 – o	80 – h
56 – w	77 – a	38 – l	70 – e
61 – b			

22 +39 61	86 –9	37 +56	52 +28
b			

48 +45	93 –67	84 –28	55 +15	70 –32	87 +3

Match the circus elephants with their blankets.

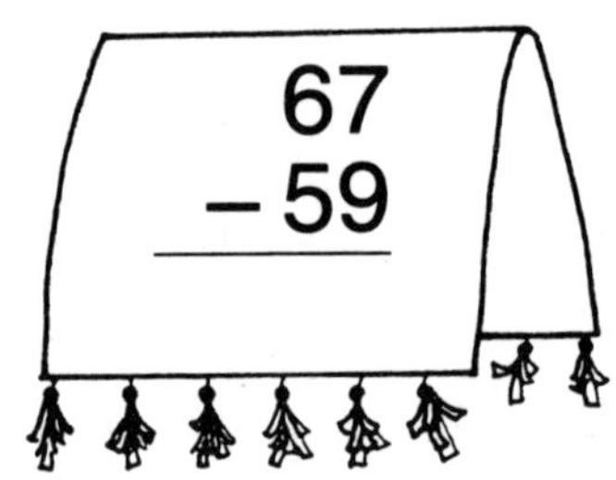

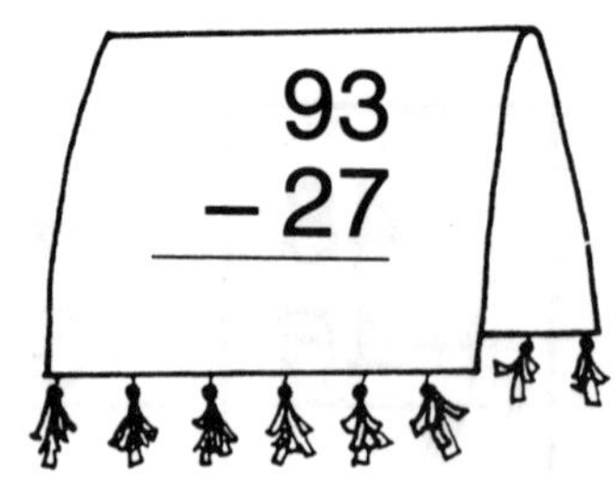

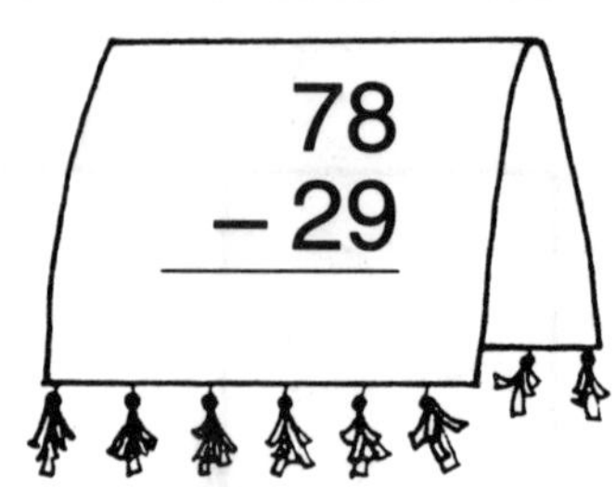

Color the **18**s red to find the elephant that ate the peanuts.

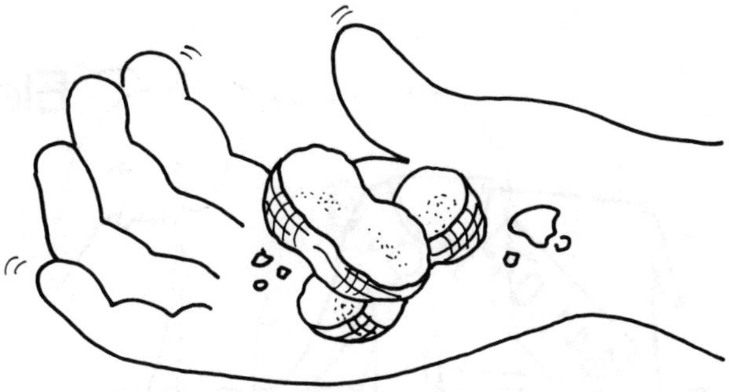

44 − 26	21 − 5	63 − 49	98 − 79	76 − 38	97 − 24
72 − 39	70 − 52	53 − 35	85 − 65	47 − 27	23 − 16
82 − 67	32 − 26	35 − 19	97 − 79	46 − 27	60 − 46
40 − 29	51 − 27	38 − 19	50 − 25	32 − 14	81 − 63
64 − 37	58 − 36	78 − 59	25 − 17	84 − 62	65 − 47

Elephants for Sale

How much will we cost?

$4.65 + 6.10	$5.23 + 4.65	$12.25 + 22.22	$41.04 + 14.30
$41.04 22.22 + 5.23	$4.65 5.23 + 6.10	$12.25 14.30 + 41.04	$22.22 13.30 + 12.25

How much more do we cost?

$8.69 – 6.10	$8.69 – 4.69	$37.58 – 22.22	$37.58 – 14.30

What do you get if you pour boiling water down a rabbit hole?

A	1,090,189	R	1,392,494
B	6,420,403	S	4,988,740
C	9,999,999	T	8,980,028
H	2,432,107	U	1,935,263
N	4,376,909	Y	1,000,000
O	1,777,777		

5,387,009 −4,296,820

7,000,000 −4,567,893	7,975,240 −6,197,463	4,286,300 +4,693,728

7,396,587 +2,603,412	6,196,247 −4,803,753	1,479,398 + 298,379	6,698,327 −1,709,587	2,494,370 +2,494,370

3,792,864 +2,627,539	8,770,359 −6,835,096	9,070,518 −4,693,609	2,288,925 +2,087,984	196,247 +803,753

Monkeys and Gorillas

1. The zoo has 17 monkeys and 9 gorillas. How many more monkeys than gorillas are there?

2. The monkeys play 7 hours a day. If they play 3 hours before lunch, how many hours do they play after lunch?

3. The children saw a tree full of monkeys. 18 monkeys were in the tree. 5 monkeys climbed out of the tree. How many are still in the tree?

4. It took the gorilla 14 minutes to eat his lunch. It took the monkeys 9 minutes. How much longer did the gorilla need?

5. Mrs. Baker needs to buy 19 boxes of food for the animals. She bought 4 boxes at one store and 5 boxes of food in the next store. How many more boxes of food does she need?

6. Koko has been in the zoo for 15 years. Bobo has been in the zoo for 6 years. How much longer has Koko been in the zoo?

Find the mistakes and correct them.

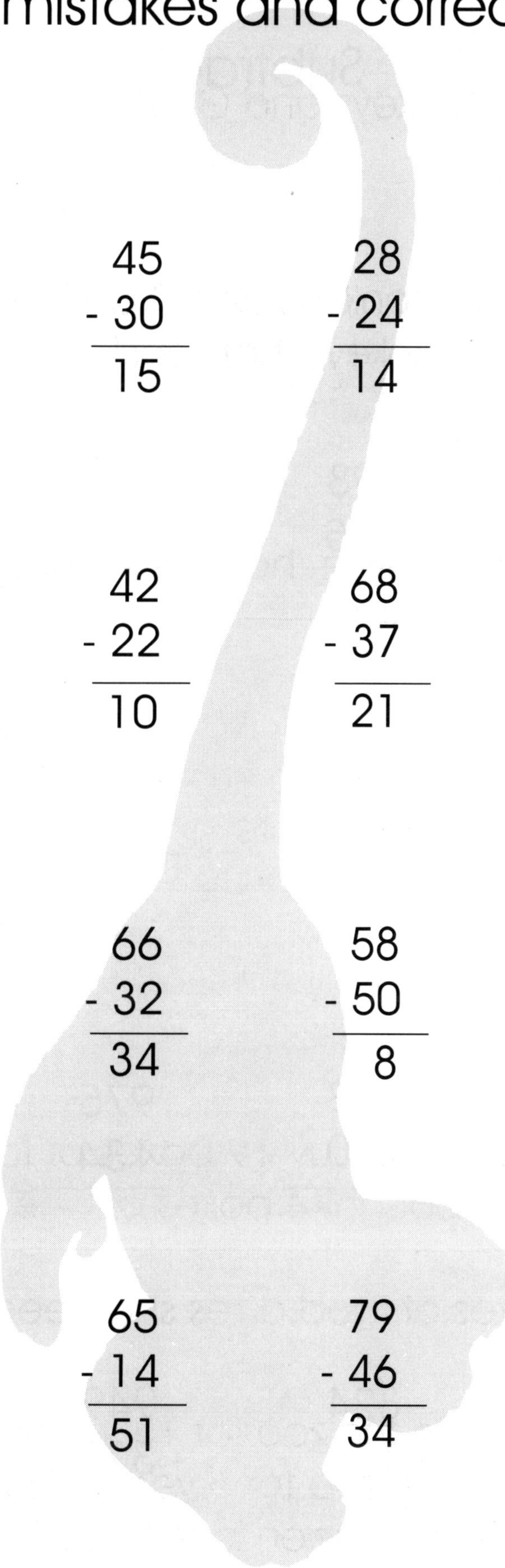

68	45	28	34
- 12	- 30	- 24	- 11
56	15	14	23
27	42	68	45
- 10	- 22	- 37	- 13
17	10	21	32
79	66	58	73
- 45	- 32	- 50	- 41
53	34	8	22
86	65	79	59
- 55	- 14	- 46	- 37
31	51	34	25

Subtract

837 - 614 223	398 - 235	254 - 122	473 - 351
256 - 142	638 - 231	356 - 214	189 - 123
643 - 331	532 - 210	975 - 134	837 - 213
584 - 121	324 - 102	949 - 325	613 - 312

How Many?

Use your ruler and measuring tape to help you answer these questions.

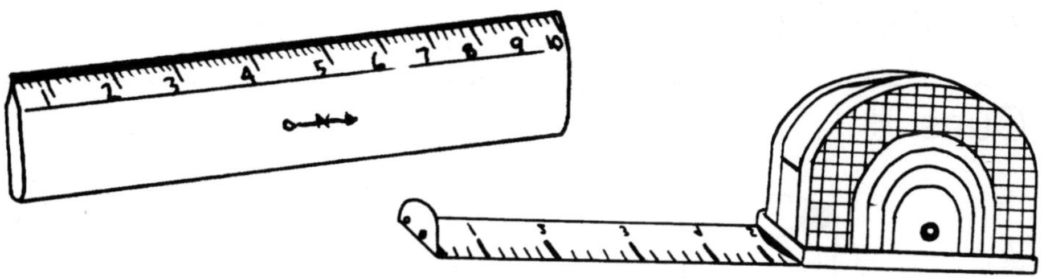

Match:

inches in a yard	100
centimeters in a meter	12
feet in a yard	36
inches in a foot	3

Answer these word problems.

1. How many inches in 2 feet? ____________________

2. How many feet in 5 yards? ____________________

3. How many centimeters in 3 meters? ____________________

4. How many inches in 2 yards? ____________________

Parents: Your child will need a set of measuring spoons and cups to do this page. Have paper towels handy since water may be spilled.

Measuring Cups and Spoons

Use water and measuring cups and spoons to help you figure out the answers to these questions.

Measuring Cups

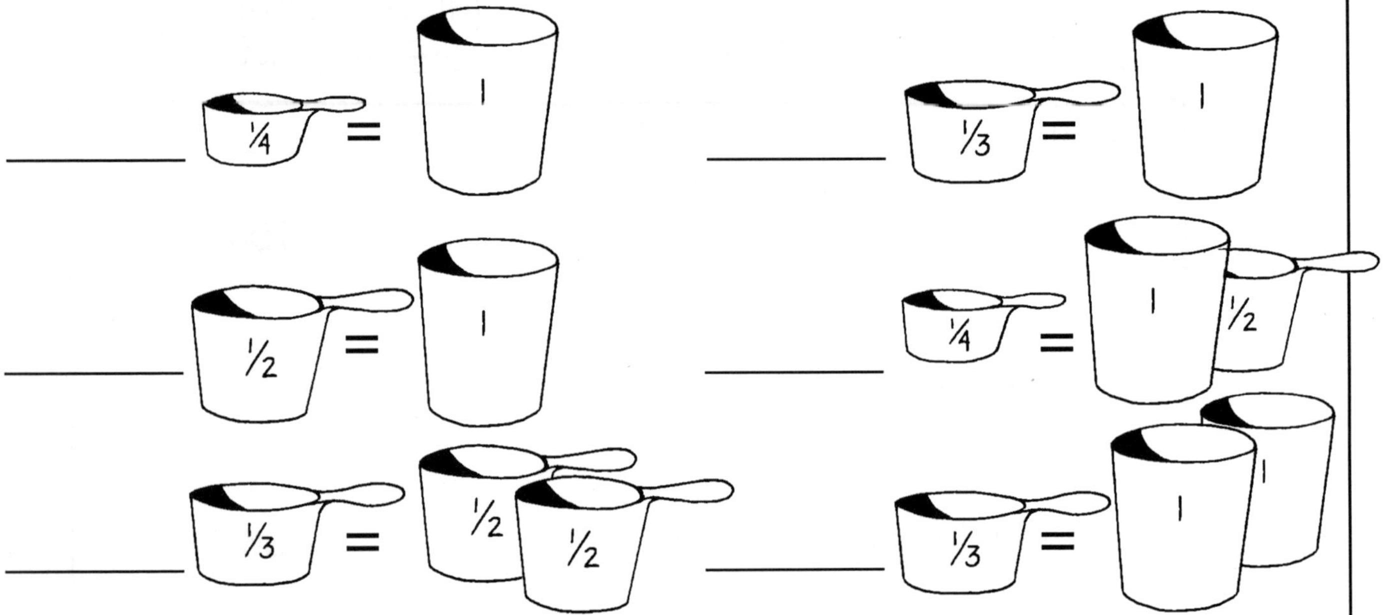

Measuring Spoons

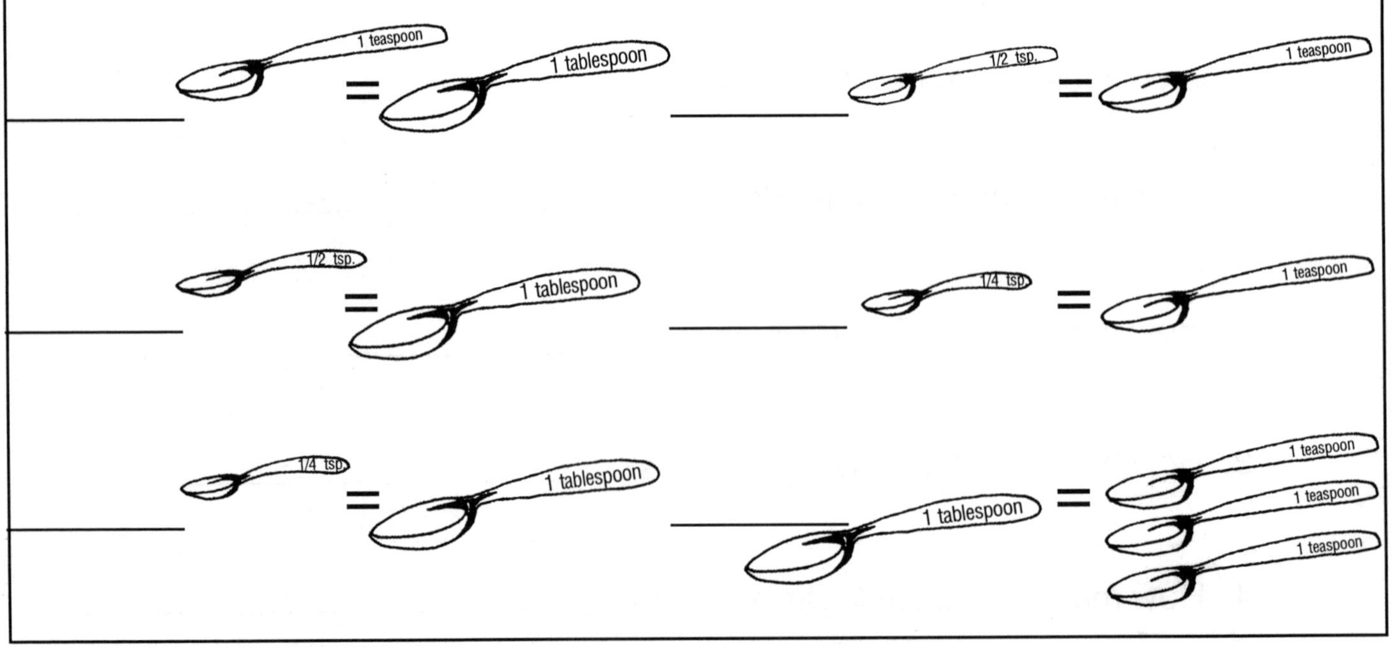

Riddles

Paste the answer under each riddle.

I am used to measure large things like lumber.

paste

I am used to measure flour when you bake.

paste

I am used to measure pieces of cloth.

paste

I am used when you want to draw a straight line.

paste

I am a container for milk and juice.

paste

I am used to measure little things like salt and spices.

paste

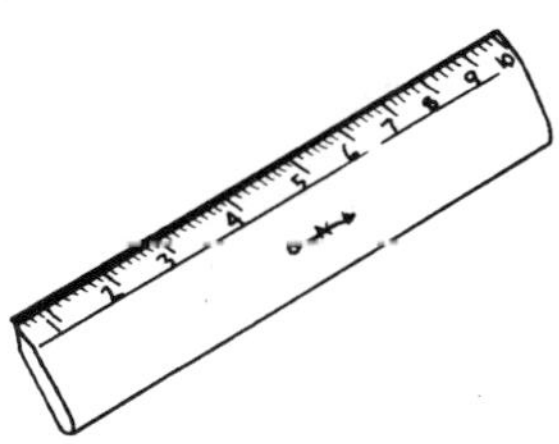

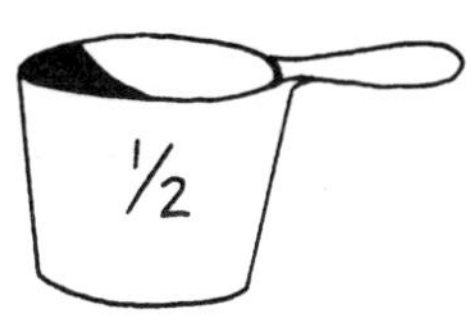

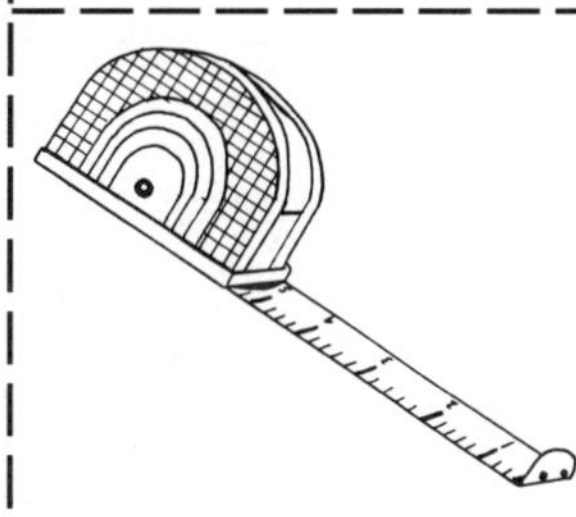

Parents: Encourage your child to draw out the problems to find the answers.

Find the Answers

Five people are going on a beach picnic.

Answer these problems to find out food and drink will they need.

1.

Work Space

There are 3 glasses of water in each liter bottle. How many liters do you need if 3 people drink 2 glasses of water and 1 person drinks 3 glasses of water?

2.

It takes 2 tablespoons of peanut butter and 3 tablespoons of jelly to make a sandwich.

How much peanut butter do you need for 5 sandwiches?

How much jelly do you need for 5 sandwiches?

3.

Each person will drink two cups of lemonade. How many cups do you need?

How Heavy Is It?

16 ounces = 1 pound

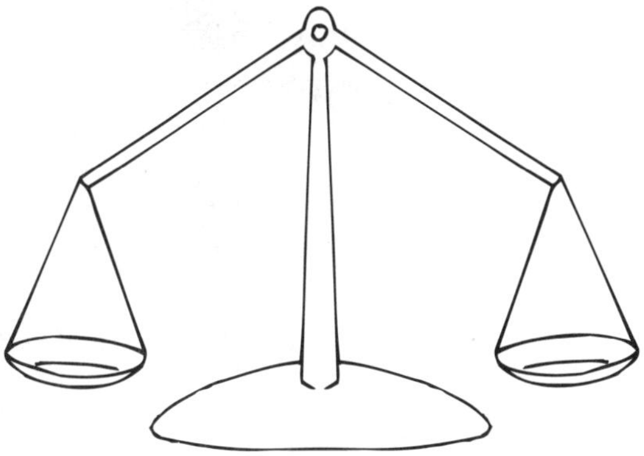

Some things you buy are measured by weight.

- packages at the post office
- dog food at the pet store
- nails at some hardware stores
- food at the supermarket

In the United States these are measured in **pounds** and **ounces**.
In most other countries they are measured in **grams**.

Look around your house and garage.

1. Find something that is measured in:

   pounds ____________________

   ounces ____________________

   grams ____________________

2. Find something that weighs about:

   5 pounds ____________________

   8 ounces ____________________

   1 pound ____________________

   16 ounces ____________________

   12 ounces ____________________

   227 grams ____________________

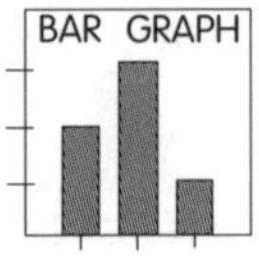

When do you go to bed?

Look at this graph.
Answer these questions.

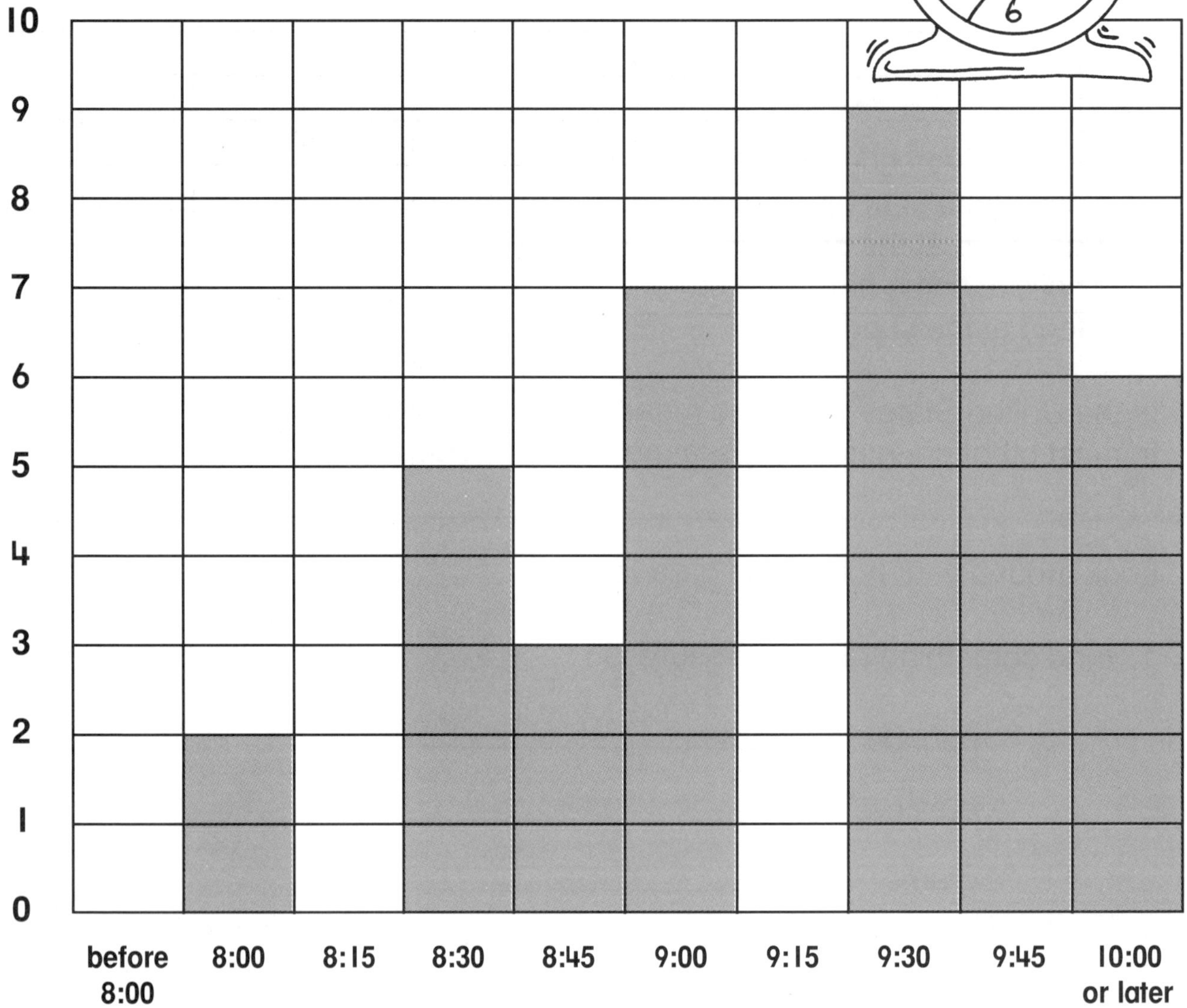

1. What is the earliest anyone went to bed? ______________

2. How many people went to bed at 10:00 or later? ______________

3. At what time did the most people go to bed? ______________

4. What time do you go to bed? ______________
Show the time on this graph. Mark your box like this: ▨

5. At what times did no one go to bed? ________________________________

Make a Bar Graph

Collect the information listed below and complete this bar graph.

Bed Time

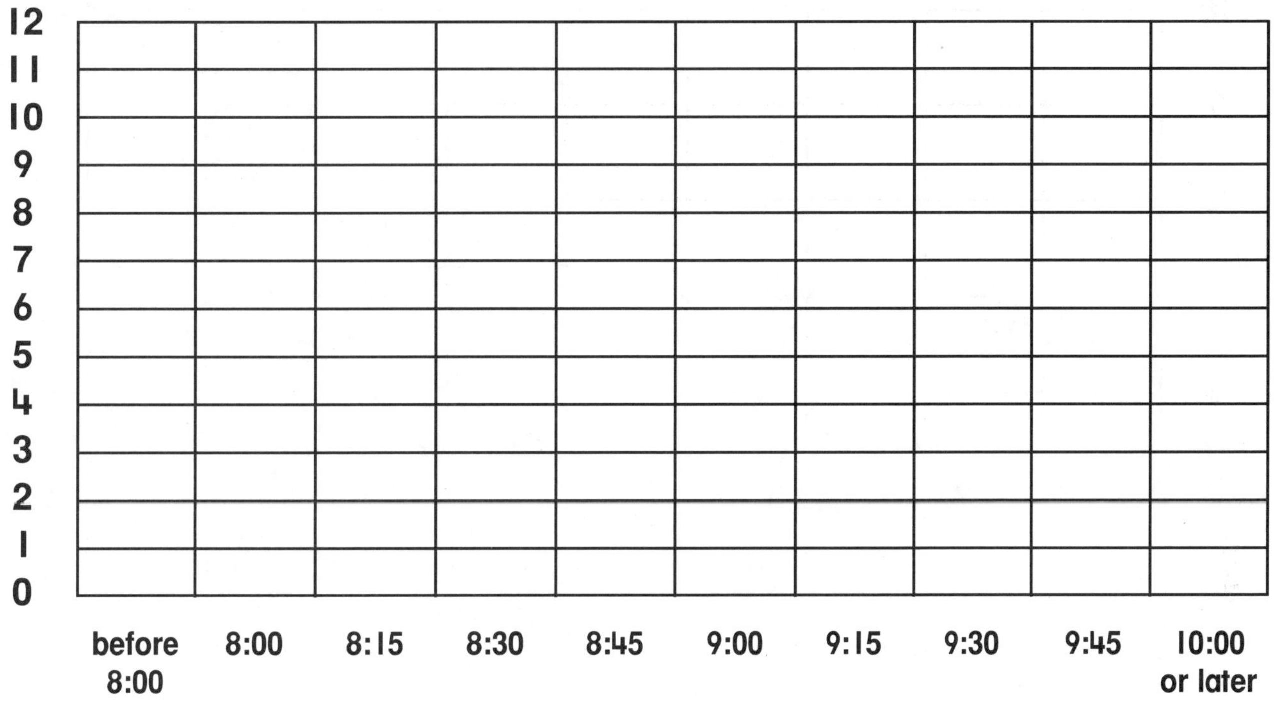

1. Collecting information and making the graph.

- **a.** Color a box on the graph to show your bed time.
- **b.** Ask each person in your family, "What time did you go to bed last night?" Color one box on the graph for each answer.
- **c.** Ask 10 more people you know the same question. Color one box on the graph for each answer.

2. Answer questions about your graph.

- **a.** What time did the most people go to bed? __________
- **b.** How many people went to bed at the same time as you? __________
- **c.** What was the latest time anyone you asked goes to bed? __________
- **d.** What was the earliest time anyone you asked goes to bed? __________

Parts of a Bicycle

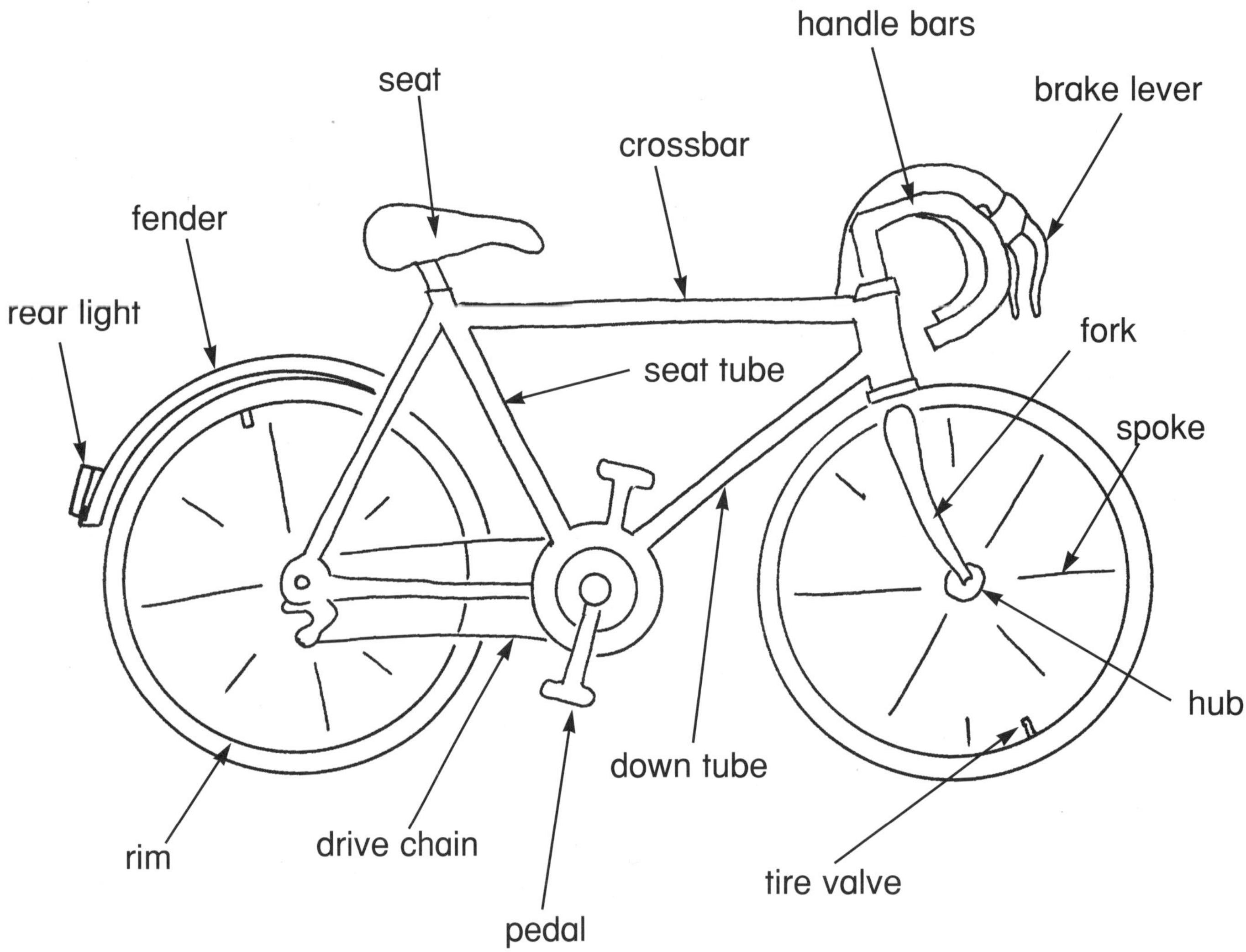

1. What do you call the part of the bicycle that holds the seat? ____________________

2. Where do you put the air in a bicycle tire? ____________________________

3. Name three parts of the bicycle wheel. ____

____

4. What is the part of a bicycle where you put your feet called?

____

5. The rear light is on the ____________________________

Big Sale!

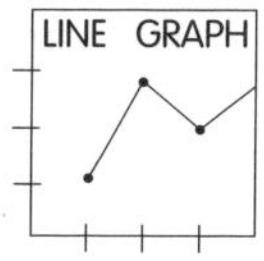

Read the graph to help you answer the questions below.

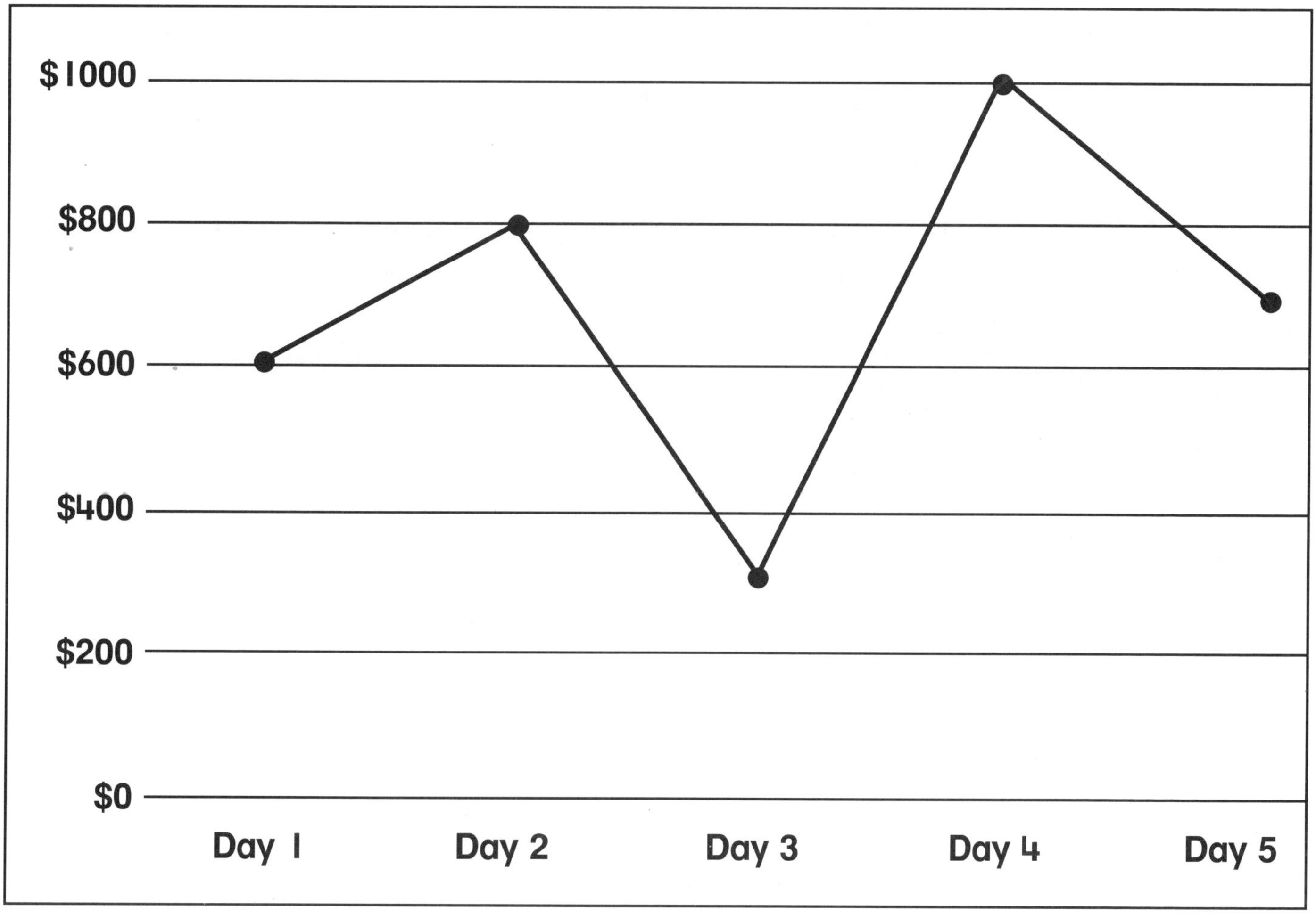

1. How much was sold on _________?

    Day 3 ____________

    Day 1 ____________

    Day 5 ____________

2. On what day was the greatest amount sold? ____________

3. On what day was the smallest amount sold? ____________

4. How much more was sold on Day 4 than on Day 1? ____________

Bonus

5. How much was sold all together? ____________

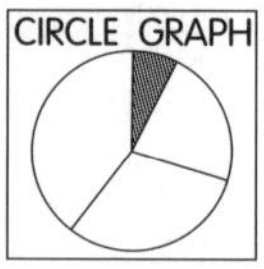

Make a Circle Graph

I asked 24 children "How do you get to school?
I collected these answers:

walk—12	ride bike—3
ride bus—6	ride in a car—3

Record this information on the circle graph below. Color in the part of the story that shows the number for each way the children came to school.

walk—red bike—blue bus—yellow car—green

Hint: Each piece of the circle stands for 3 children.

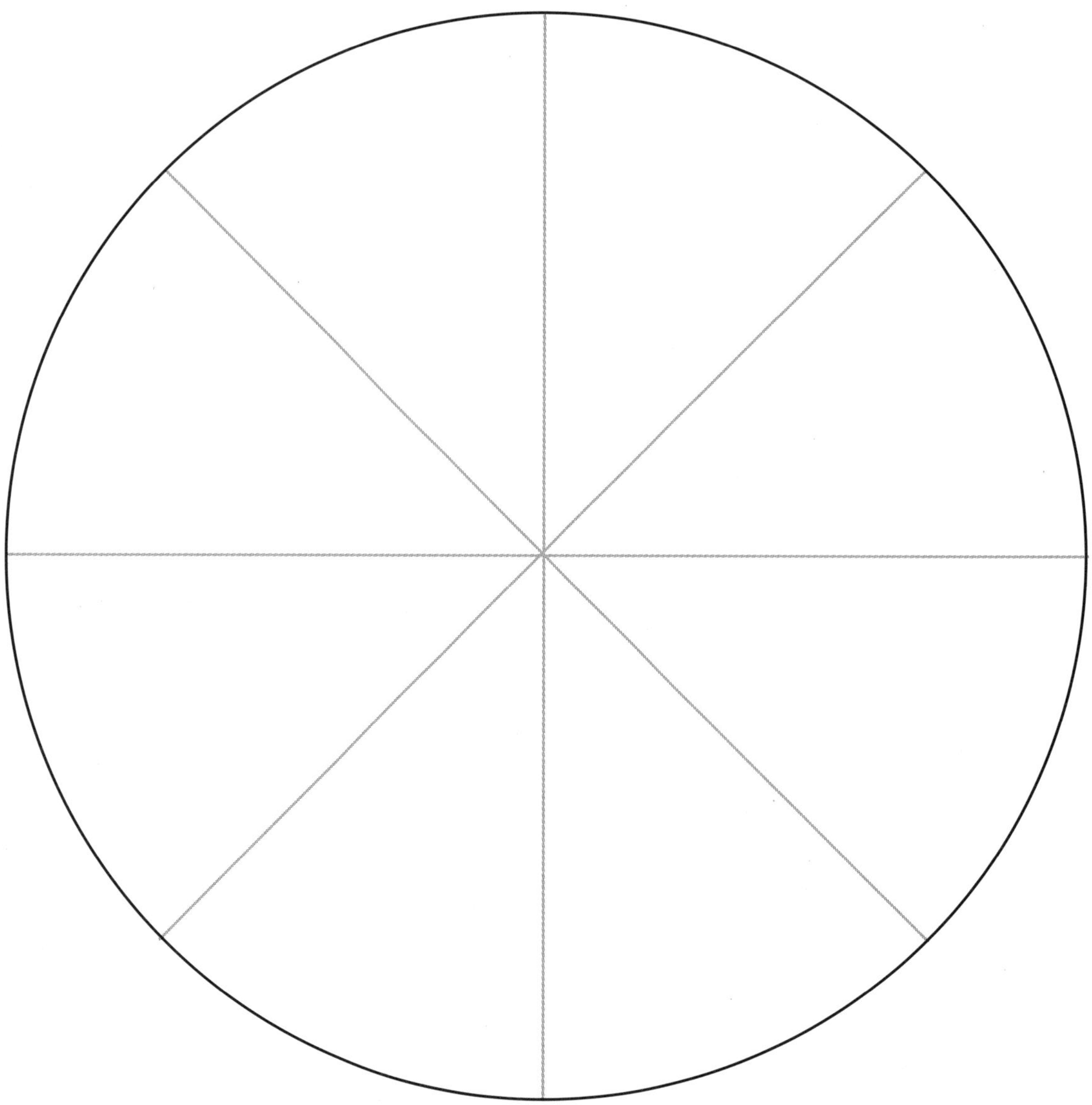

I can see 15 people on the street.
5 have open umbrellas.
7 have closed umbrellas.

How many don't have an umbrella?

I can solve the problem.

- Look at the pattern.
- How many flowers will go in the next vase?

I can solve the problem.

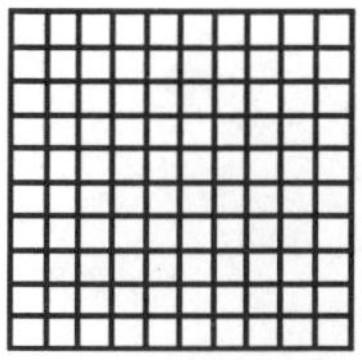

= 100
little blocks

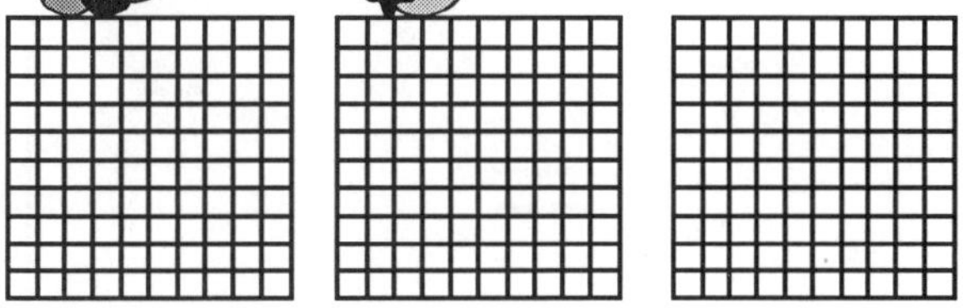

= ________

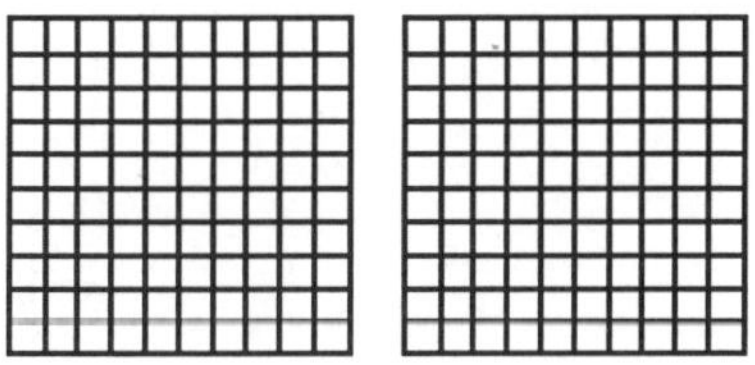

= ________

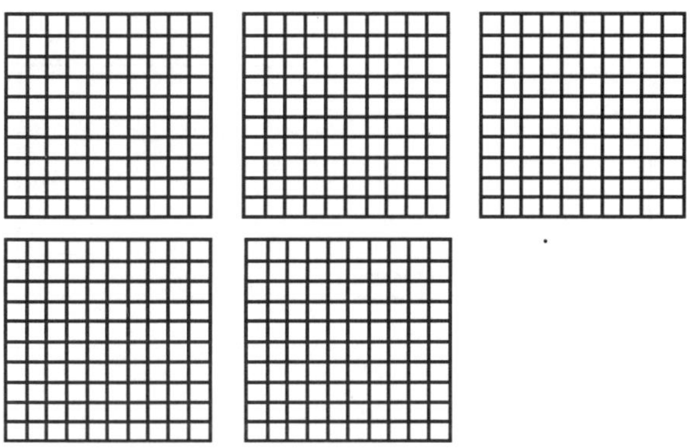

= ________

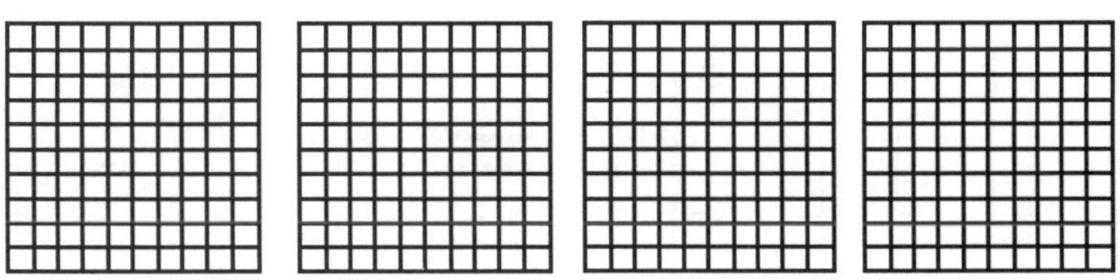

= ________

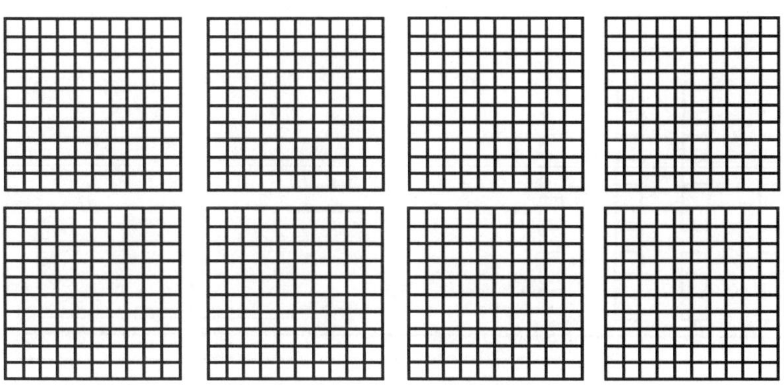

= ________

More than 100.

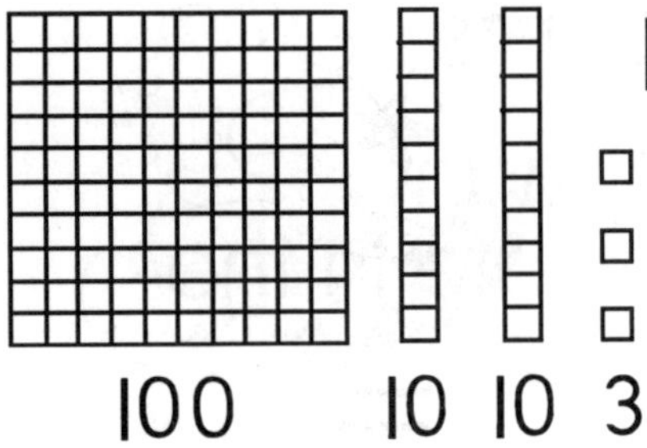

= 123

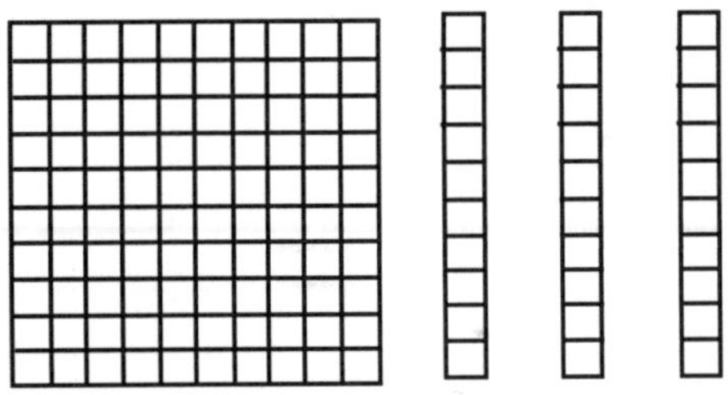

= ______

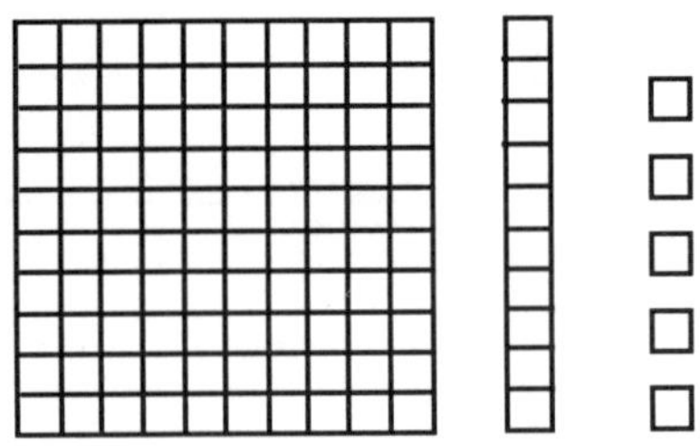

= ______

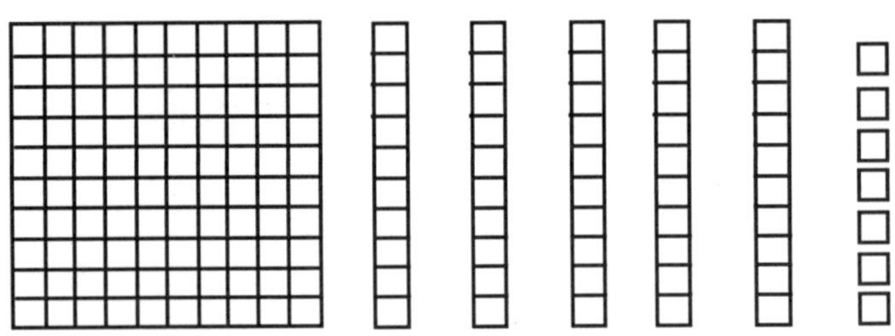

= ______

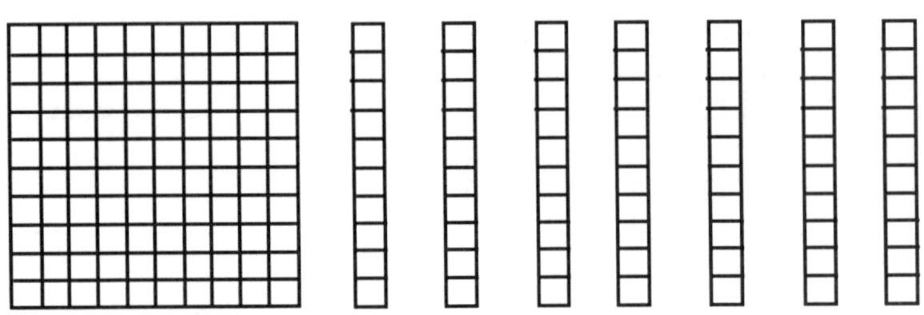

= ______

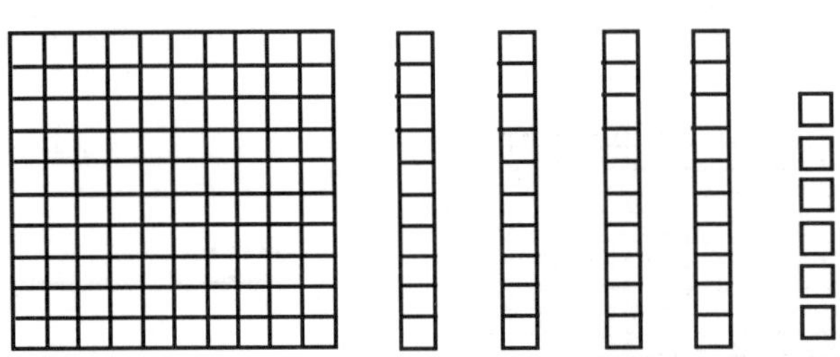

= ______

Counting BIG Amounts

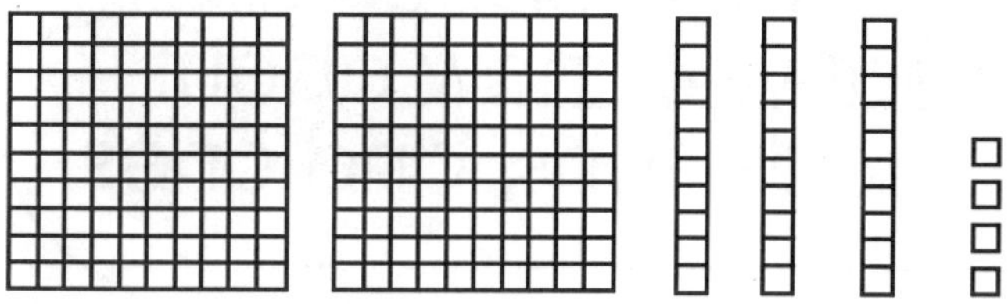

= 234 ______

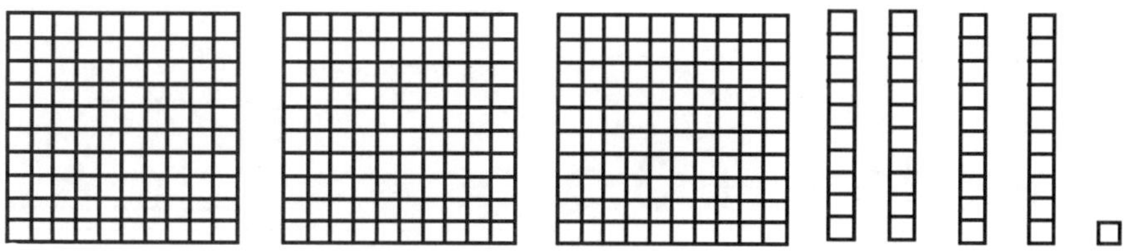

= ______

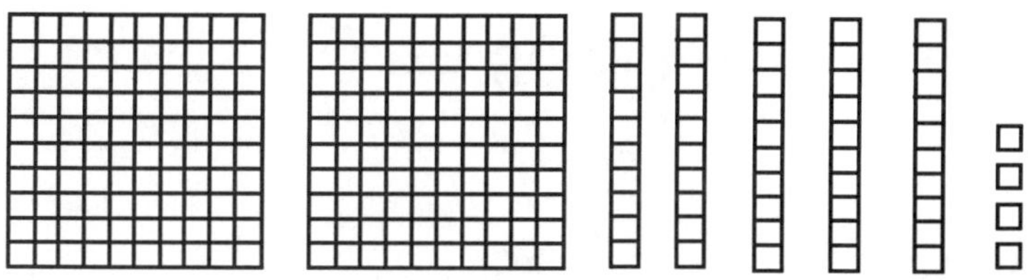

= ______

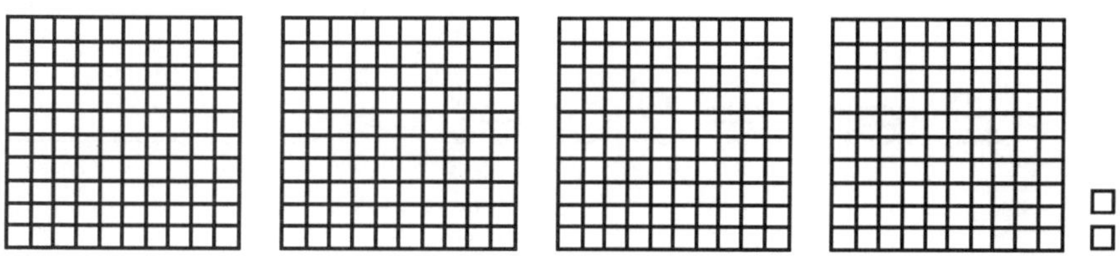

= ______

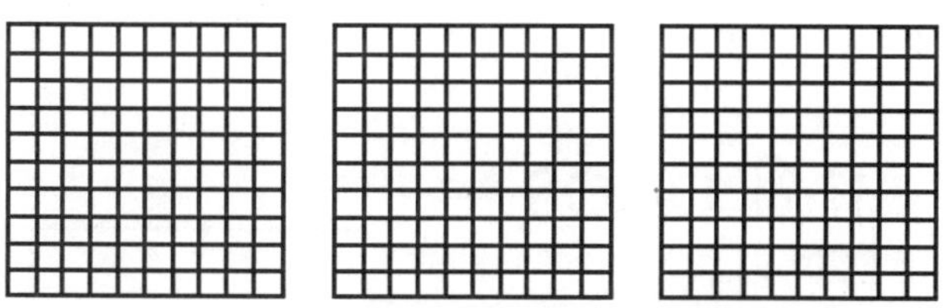

= ______

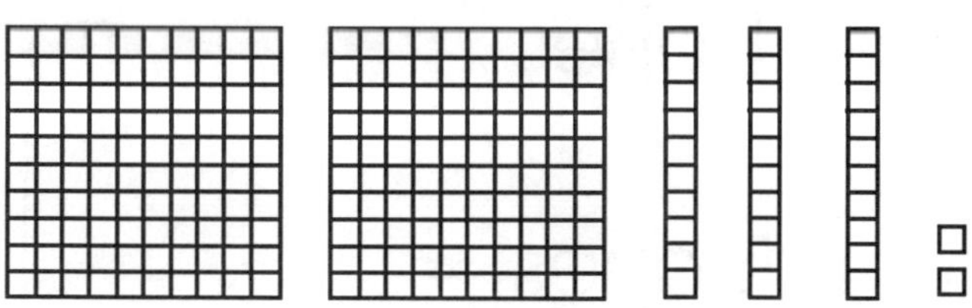

= ______

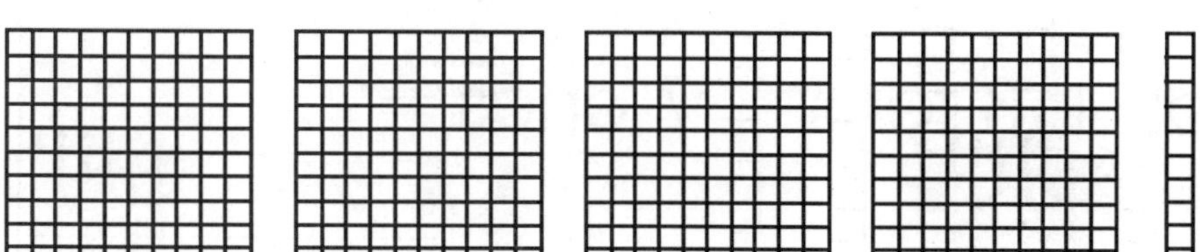

= ______

5 hundreds + 6 tens + 3 ones = 563

6 hundreds + 8 tens + 6 ones = ________

7 hundreds + 4 tens + 9 ones = ________

8 hundreds + 2 tens + 2 ones = ________

9 hundreds + 6 tens + 4 ones = ________

8 hundreds + 1 tens + 8 ones = ________

7 hundreds + 7 tens + 1 ones = ________

6 hundreds + 9 tens + 5 ones = ________

5 hundreds + 3 tens + 7 ones = ________

REVIEW 3

Skills: • identifies numbers both in word and numeral form
• identifies odd and even numbers

Examples: these numbers are odd - 1, 3, 17, 25
these numbers are even - 4, 8, 16, 26

Directions: Circle or write the correct answer for each question.

1. Which number is three hundred twenty-seven?

a. 723 b. 327 c. 3270 d. 237

2. Which number is nine thousand forty?

a. 904 b. 9004 c. 940 d. 9040

3. Which number is eighty-four thousand two hundred?

a. 80,200 b. 84,200 c. 8420 d. 84.020

4. Draw lines to match each number with its word name:

630	sixty-three
6300	six hundred thirty
63	six thousand three hundred
603	six hundred three

5. Circle the odd numbers:

95 34 28 43 81 18

6. Circle the even numbers:

98 82 26 35 39 64

3 REVIEW

Skills: • identifies ordinal numbers
• understands expanded form
• uses (greater than) (less than) symbols correctly

Examples: ordinal numbers - first, sixth etc.
expanded form - 467 = 400 + 60 + 7
63 > 60, 60 < 63

Directions: Choose or write the correct answers to each question.

1. Rewrite in the correct order:

Fourth ____________________

Sixth ____________________

First ____________________

Second ____________________

Third ____________________

Fifth ____________________

2. Rewrite each number in expanded form:
For example: 562 = 500+60+2

78 ____________________

245 ____________________

3. Read the expanded form; then write the number
For example: 200+30+7 = 237

800 + 20 + 6 ____________________

100 + 90 + 9 ____________________

4. Put > or < in each circle:

59 ◯ 61 304 ◯ 314

145 ◯ 135 5972 ◯ 5952

REVIEW 3

Skills: • shows understanding of number order
• can repeat a number pattern
• can use +, -, x signs to make true number statements

Directions: Write the correct answers to each question.

1. Rewrite the numbers in the correct order from smallest to largest:

839 29 391 48 85 17 8

_____ _____ _____ _____ _____ _____ _____

smallest largest

2. Continue the pattern:

4 7 10 _____ _____

55 53 51 49 47 _____ _____

3, 4, 5 13, 14, 15 23, 24, 25 _____, _____, _____

3. Write the number that is one before and one after the numbers shown:

_____ 420 _____ _____ 399 _____

_____ 1050 _____ _____ 7357 _____

_____ 94,671 _____ _____ 26,519 _____

4. Put +, -, or x in each circle to make the number statement true:

3 ◯ 5 = 15 17 ◯ 9 = 8

6 ◯ 4 ◯ 8 = 18 6 ◯ 4 ◯ 8 = 2

3 REVIEW

Skills: • counts by 2s, 5s and 10s to 100
• understands place value

Directions: Write the correct answer to each question.

1. Write the numbers to count by 2s:

22 ______ ______ ______ ______ ______ 34

2. Write the numbers to count by 5s:

50 ______ ______ ______ ______ ______ 80

3. Write the numbers to count by 10s:

40 ______ ______ ______ ______ ______ 100

4. Round each number to the nearest 10:

58 = almost _____ 61 = almost _____ 29 = almost _____

5. Look carefully at each number. Circle the number in the ones place; put an X on the number in the hundreds place.

106 948 1391 4752

6. Look carefully at each number. Put a box around the number in the tens place; put a line under the number in the thousands place.

2350 9675 4000 8888

7. Write the numbers for each of these:

4 tens and 8 ones = ____________________

4 hundreds, 5 tens and 2 ones = ____________________

8 thousands, 7 hundreds, 0 tens and 6 ones = ____________________

4 thousands, 0 hundreds, 3 tens = ____________________

REVIEW 3

Skills: • adds and subtracts single digit numbers in a series
• adds and subtracts multi-digit numbers without borrowing and carrying

Directions: Do not use your calculator for this page.

1. Add or subtract. Watch the signs!

$3 + 4 + 7 =$ $\qquad$ $5 - 2 + 8 =$

$3 + 5 - 7 + 9 - 2 =$ $\qquad$ $19 - 3 + 5 - 1 + 4 =$

2.

49	30	384	994	4651
- 21	+ 22	- 103	- 4	+ 1326

3.

845	75	829	104	9805
- 31	+ 22	- 726	+ 14	-2300

4. Study the picture. Then answer the questions.

TOY STORE

How much will you need to buy the ball and the truck? ____________

How much more does the doll cost than the yo-yo? ____________

3 REVIEW

Skill: solves one-step word problems requiring addition and subtraction without regrouping (carrying and borrowing)

Directions: Read the information. Then work each problem in the space provided. Write your answer on the line. Be sure to label your answer, for example: 8 big dogs. 4 small dogs. How many dogs? 8 + 4 = 12 12 dogs The four third grade classes are going on a trip to a park. Two classes have 23 children and two classes have 21 children.	**1.** How many children will go on the trip to the park? __________
2. A school bus holds 50 children. How many buses will be needed? __________	**3.** It takes 24 minutes to go from the school to the park. How long is a round trip? __________
4. At the park 24 children want to go for a hike. How many will stay behind? __________	**5.** The park has 7 swings and a merry-go-round that holds 12 children. Can Ms. Jones class of 21 students use the equipment at the same time? __________

Skill: adds and subtracts multi-digit numbers with regrouping (carrying and borrowing)

Directions: Add or subtract. Do not use your calculator. Watch the signs!

1.	57 + 35	86 + 47	325 - 69	700 - 450	588 + 281
2.	836 - 379	5066 + 1789	4290 + 3876	9429 - 5683	600 - 151

3. Study the picture. Then answer the questions.

At the Movies

Children under 12	$2.75
Adults	$6.50
Senior Citizen (62 & up)	$4.25
Silver Screen Passes	$5.00

How much for 2 adults to see a movie? ________________

How much for Grandma (age 67) to take Sarah (age 9)? ________________

3 REVIEW

Skill: solves one-step word problems requiring addition and subtraction with regrouping (carrying and borrowing)

Directions: Use the information on the map to work each problem. Do not use a calculator. Write and work each problem in the space provided. Write your answer on the line. Be sure to label your answer, for example:

8 big dogs. 4 small dogs. How many dogs?
8 + 4 = 12 <u>12 dogs</u>

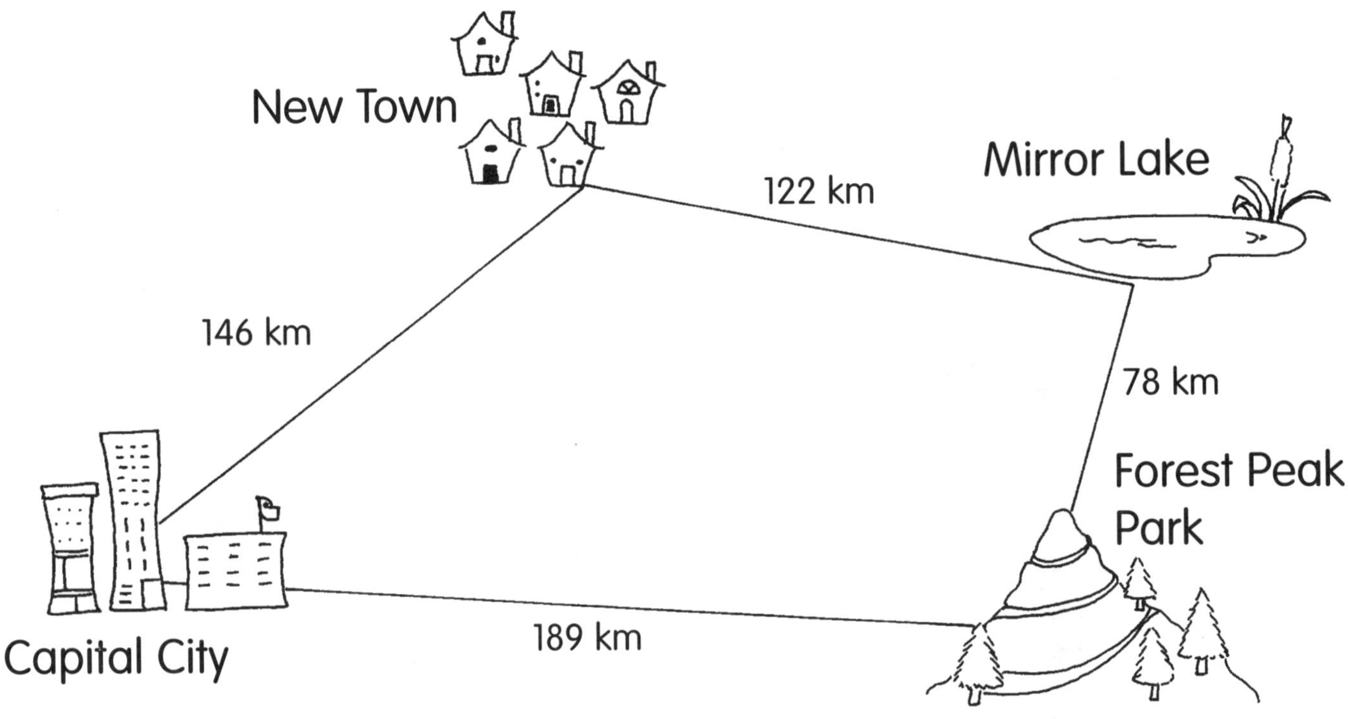

1. Mr. Diaz delivers packages. Every day he drives from Capital City to New Town. Then he drives on to Mirror Lake, Forest Peak Park and finally back to Capital City. How far does he drive every day? ____________

2. Jose and Tina live in Capital City. They want to drive to Mirror Lake. Which way is shorter - through New Town or through Forest Peak Park? ____________

3. Ms. Lee has enough gas in her car to drive 250 km. Can she get from New Town to Forest Peak Park without stopping to refuel? ____________

3 REVIEW

Skills: knows multiplication and division facts through the 5's

Directions: Do not use your calculator for this page.

1. Multiply

4 x 3 =	5 x 2 =	6 x 4 =
7 x 5 =	4 x 4 =	2 x 6 =

2. Divide

30 ÷ 5 =	18 ÷ 2 =	24 ÷ 4 =
15 ÷ 3 =	20 ÷ 5 =	9 ÷ 3 =

3. Multiply

$\begin{array}{r} 6 \\ \underline{\times 3} \end{array}$ $\quad$ $\begin{array}{r} 5 \\ \underline{\times 5} \end{array}$ $\quad$ $\begin{array}{r} 9 \\ \underline{\times 4} \end{array}$ $\quad$ $\begin{array}{r} 8 \\ \underline{\times 3} \end{array}$ $\quad$ $\begin{array}{r} 3 \\ \underline{\times 3} \end{array}$

4. Divide

$2\overline{)14}$ $\quad$ $3\overline{)18}$ $\quad$ $4\overline{)12}$ $\quad$ $5\overline{)45}$ $\quad$ $4\overline{)36}$ $\quad$ $3\overline{)21}$

3 REVIEW

Skills: • understands basic properties of multiplication
• understands the relationship between multiplication and division

Directions: Solve the problems without using a calculator.

1. 2 x 3 = ☐ x 2 6 x 7 = 7 x ☐	**2.** (3 x 4) x 2 = 24 3 x (4 x 2) = ☐
3. 2 x 1 = ☐ 5 x 1 = ☐ 1 x 6 = ☐ Any number multiplied by 1 = ____________	**4.** 4 x 0 = ☐ 9 x 0 = ☐ 0 x 5 = ☐ Any number multiplied by 0 = ☐

5. 9 x 2 = 18

18 ÷ 2 = ☐

18 ÷ 9 = ☐

6 x 5 = 30

30 ÷ 5 = ☐

30 ÷ 6 = ☐

Skill: understands whether to use multiplication or division to solve a problem

3 REVIEW

Directions: Circle the operation you would use to solve each problem.

1. Samantha, Carol and Andy have 4 pieces of gum each.
How much gum do they have altogether?

a. multiplication   b. division

2. There are 6 cards with 5 pictures on each.
How many pictures are there?

a. multiplication   b. division

3. Each package contains 4 chocolate candies.
How many packages are needed to give one candy to each of 28 students in the class?

a. multiplication   b. division

4. I have 45 photos to put in my album. I can fit 5 photos on a page.
How many pages will I use up?

a. multiplication   b. division

5. Mr. Williams buys stickers that come 6 on a sheet. He has 3 sheets.
How many stickers can he give out?

a. multiplication   b. division

3 REVIEW

Skill: uses multiplication and division facts through the 5's to solve word problems

Directions: Write and solve each problem in the space provided. Do not use a calculator. Write your answer on the line. Be sure to label your answer. For example: Bob, Carlos, and Frank each have 5 marbles. How many marbles do they have together? 3 x 5 = 15 <u>15 marbles</u>	**1.** Bunches of bananas have 4, 5, or 6 bananas. Todd bought 3 bunches. What is the smallest number of bananas he might have? __________
2. What is the largest number of bananas Todd might have? __________	**3.** Todd just loves bananas. He eats three bananas every day. How long will 18 bananas last him? __________
4. Bananas are on sale for 9¢ a pound. How much will 3 pounds of bananas cost? __________	**5.** Todd has 45¢. How many pounds of bananas at 9¢ a pound can he buy? __________

REVIEW 3

Skill: uses multiplication and division facts through the 5s to solve problems

Directions: Do not use a calculator for this page.

Complete the tables: How much will it cost to buy?

	 2 cents each	 3 cents each	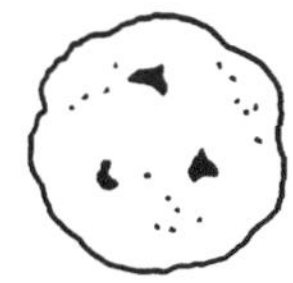 4 cents each	 5 cents each
1 costs	2 cents			5 cents
2 costs		6 cents		
3 costs	6 cents		12 cents	
4 costs				20 cents
5 costs		15 cents		

How many French fries will each person get?

There are 25 French fries		________ French fries each
There are 24 French fries		________ French fries each
There are 18 French fries		________ French fries each
There are 16 French fries		________ French fries each

3 REVIEW

Skills: • identifies plane and space (three-dimensional) figures, points, lines, and right angles
• draws lines, points and triangles

Directions: Draw a line to match each figure with its name.

Triangle
Cube
Rectangular prism
Circle
Sphere
Square

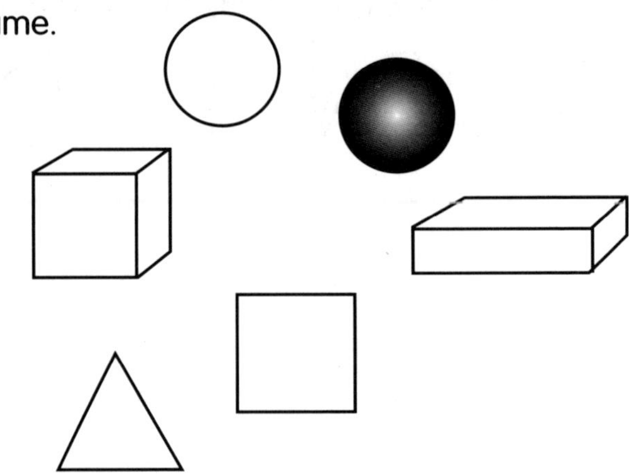

Directions: Choose or write the correct answer to the questions

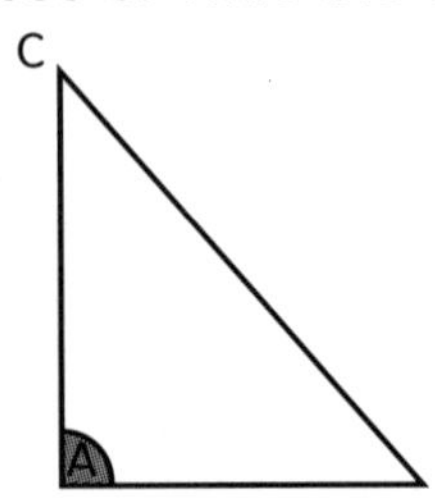

Figure 1

Figure 2

Directions: Look at Figure 1. Answer questions 1 & 2.

1. The shaded area A is a __________.
a. corner b. side c. right angle

2. A triangle __________ has 3 sides.
a. sometimes b. never c. always

Look at Figure 2. Answer questions 3 - 5.

3. AC is a __________.
a. line b. line segment c. point

4. AB is a __________.
a. line b. line segment c. point

5. B is a __________.
a. line b. line segment c. point

Skills: recognizes congruent and symmetrical figures

Examples: Figure 1 and 2 are congruent
Figure 3 and 4 are not congruent
Figure 5 is divided symmetrically
Figure 6 is not divided symmetrically

figure 1

figure 2

figure 5

figure 6

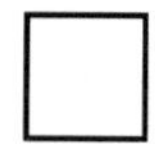
figure 3

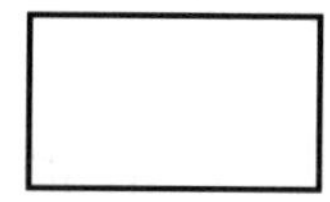
figure 4

Directions: Mark or show the correct answer.

1. Are the pairs of figures congruent?

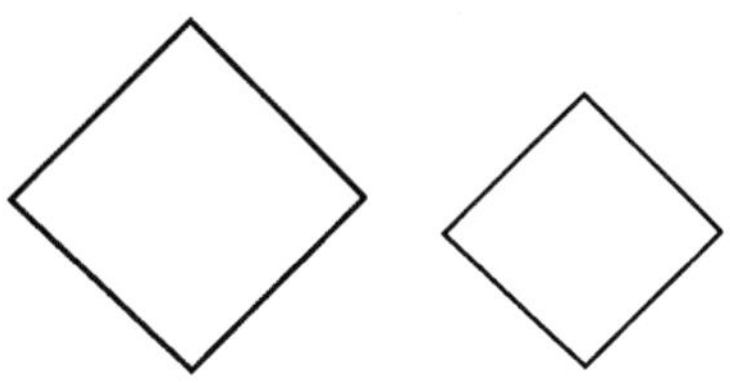
Yes No

Yes No

2. Draw a figure that is congruent to the figure shown.

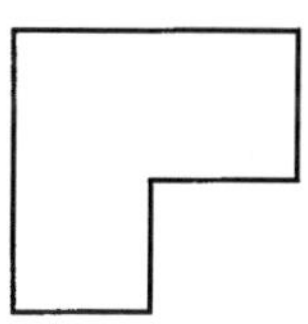

3. Is each figure divided symmetrically?

Yes No

Yes No

Yes No

4. Draw a line of symmetry on this figure.

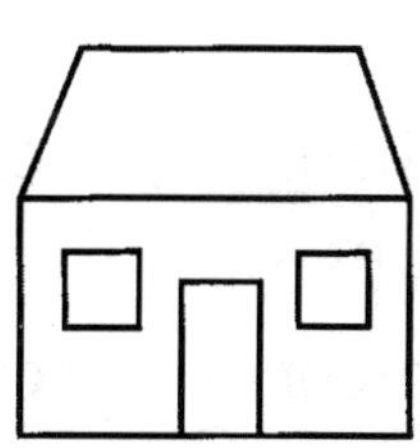

3 REVIEW

Skill: finds a location on a grid by using ordered pairs

Example: On the grid below, the ordered pair (4,6) describes the location of point A

Directions: Use the grid to answer the questions.

1. What is the ordered pair located at "B"? ______________

2. What is the ordered pair located at "C"? ______________

3. What is the ordered pair located at "E"? ______________

4. Put a dot at (6,1). Label it D.

5. Put a dot at (4,2). Label it F.

6. Put a dot at (6,3). Label it G.

7. Connect these points: BC CD DE BFE FA AG

8. What did you draw? ______________________

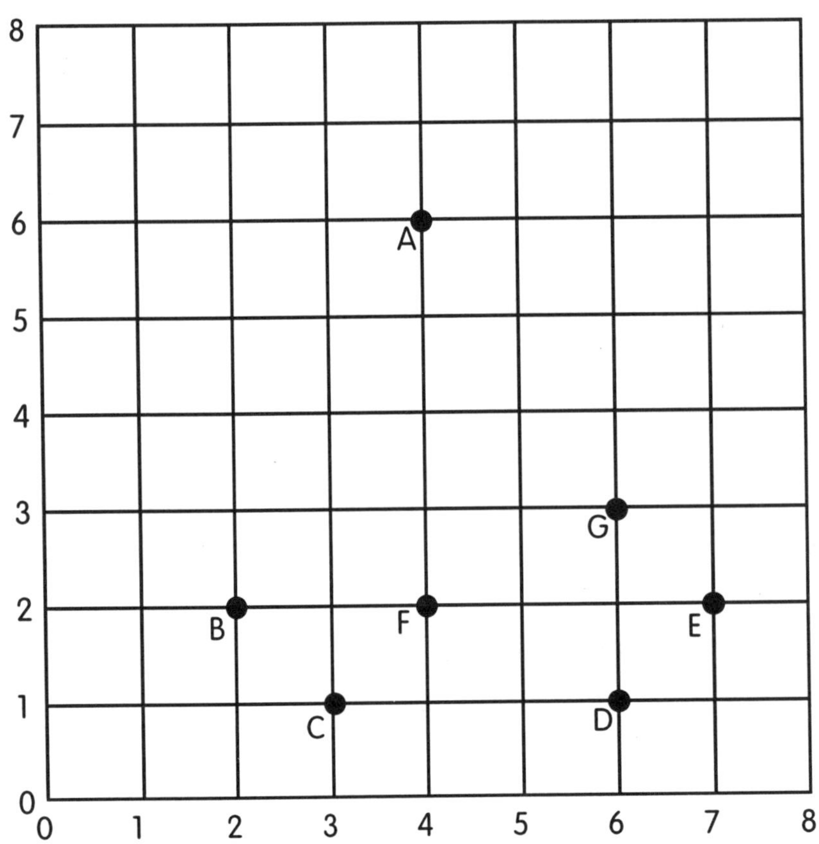

3 REVIEW

Skills: • determines the area of a figure by counting units
• determines the perimeter of a figure

Directions: Choose or write the correct answers to the questions. Study the figures. Then answer the questions.

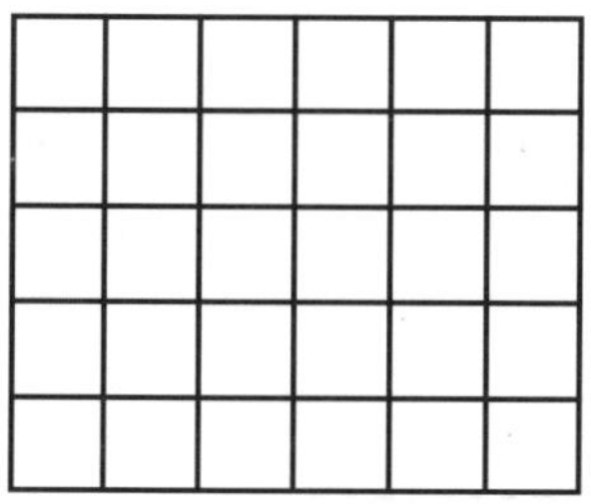

figure A

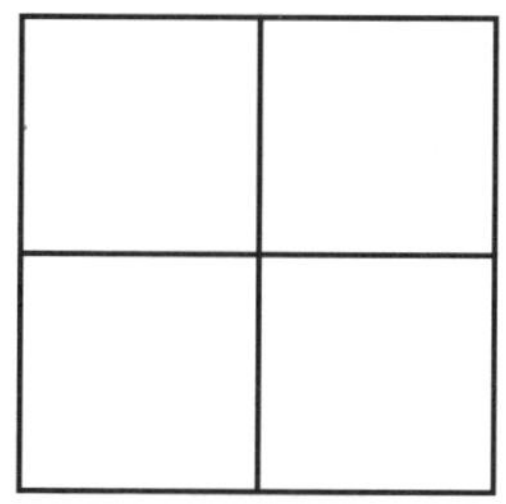

figure B

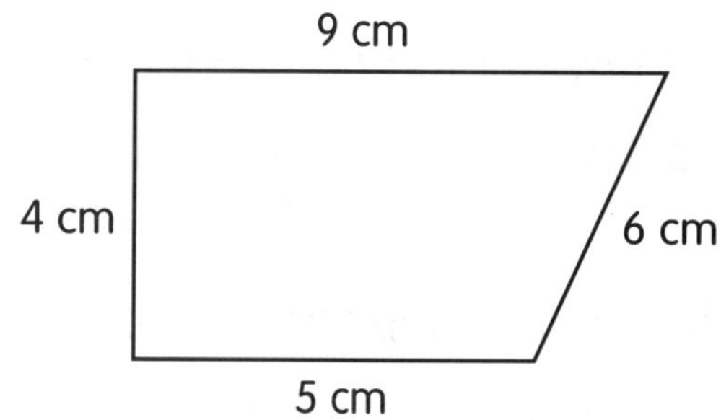

figure C

1. What is the area of Figure A?

a. 30 square units b. 60 square units c. 36 square units

2. What is the area of Figure B?

a. 8 square units b. 4 square units c. 16 square units

3. What is the perimeter of Figure C? ________________ cm

3 REVIEW

Skills: Chooses an appropriate unit to measure for a real life situation.

Directions: What unit of measurement would you use? Choose the most sensible one.

1. To tell how far it is around the earth

a. feet b. inches c. miles d. ounces

2. To tell how tall the Statue of Liberty is

a. meters b. centimeters c. grams d. kilometers

3. To measure the water in a swimming pool

a. ounces b. cups c. pints d. gallons

4. To tell how much juice to buy for the class party

a. milliliters b. liters c. kilograms d. grams

5. To tell how much a cow weighs

a. grams b. kilograms c. meters d. kilometers

6. To tell how much an apple weighs

a. ounces b. pounds c. tons d. quarts

REVIEW 3

Skills: • measures to the nearest inch or centimeter
• estimates the weight and capacity of pictured objects

Directions: Choose or write the correct answers to the questions.

Note: You will need a ruler that measures in inches and centimeters

line A   line B

1. Measure line A in centimeters. ____________cm

2. Measure line B in inches. ____________inches

3. Draw a line from each picture to <u>two</u> ways of measuring it.

5 gallons

5 kilograms

2 meters

10 pounds

19 liters

6 1/2 feet

11 pounds

1 cup

240 milliliters

3 REVIEW

Parents: Today digital clocks are everywhere. Many children do not know how to tell time on an analog clock (a clock with hands). It's important to have analog clocks at home and to make sure your child's first watch is analog, not digital.

Skills:
- recognizes time in both analog and digital form
- tells time to the minute in more than one way
- can add or subtract from the time represented on a clock

Example: Tell the time in 3 ways

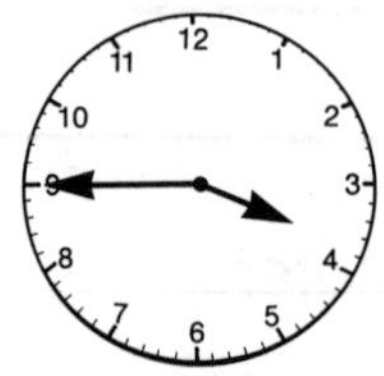

3:45
45 minutes after 3
15 minutes to 4

Directions: Choose or write the correct answers to the questions.

1. Match the clocks

12:30

10:15

8:08

2. Tell the time in 3 ways.

_____ : _____
_____ minutes after _____
_____ minutes to _____

_____ : _____
_____ minutes after _____
_____ minutes to _____

3. Look at the clock. Show a half-hour later.

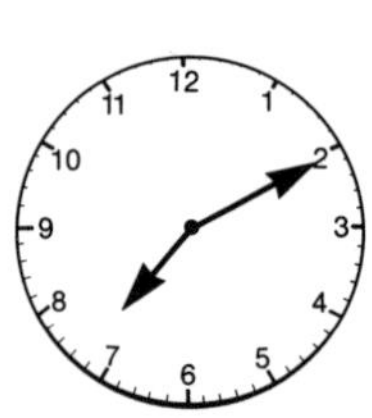

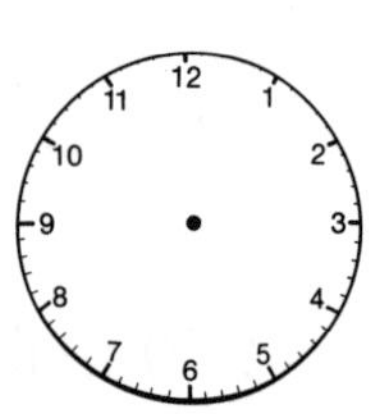

4. Look at the clock. Show 15 minutes earlier.

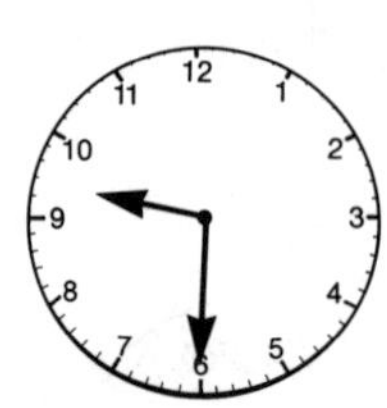

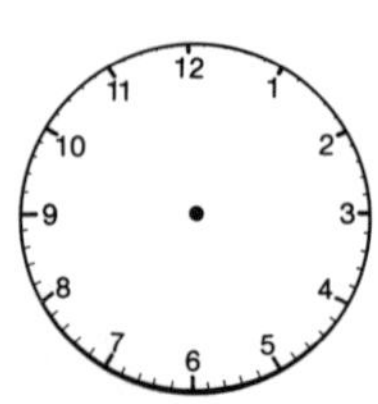

Skills: reads and interprets temperature in both Centigrade and Fahrenheit.

Directions: Write or choose the correct answers to the questions.

This is the thermometer John watched today.

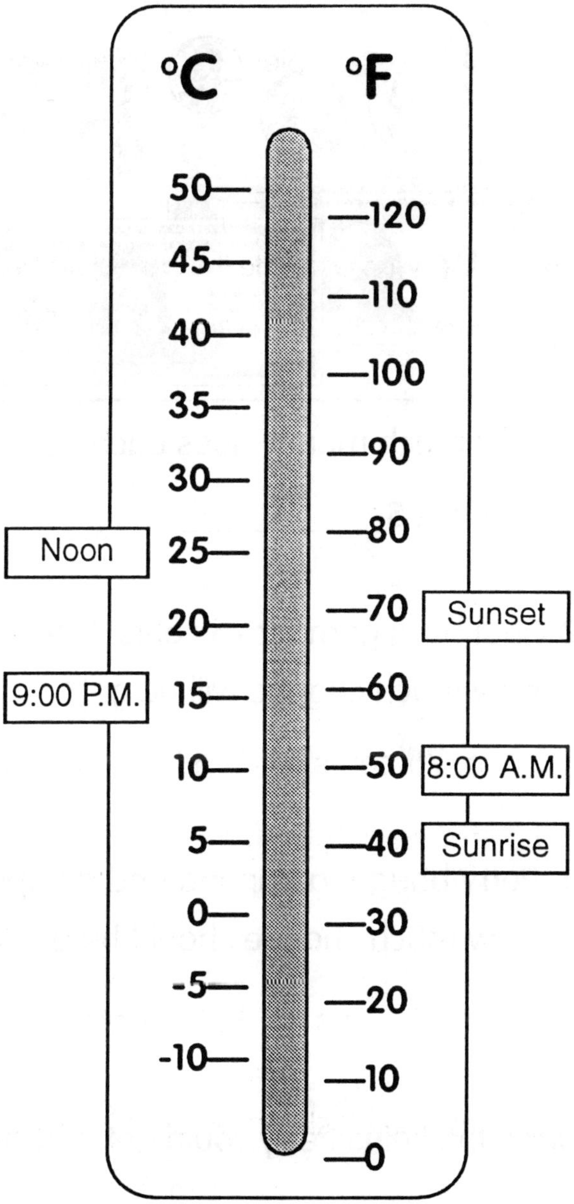

1. When John got up at 8:00 A.M. the thermometer read 50°F; What was that on the Centigrade scale? ___________°C

2. The highest temperature of the day was 30°C. Was it hotter than 90°F? ___________

3. What was the temperature in °F when the sun went down? ___________°F

4. How much cooler (°C) was it at 9:00 P.M than at noon? ___________°C

5. How much higher was the temperature at sunset than at sunrise (°F)? ___________°F

6. John went to bed at 9:00 P.M. Was it warmer or colder when he woke up at 8:00 A.M.? ___________

How many degrees Centigrade? ___________°C

3 REVIEW

Skills: • counts money in coins and bills to $10
• makes change for $1.00

Directions: Write or choose the correct answer for each question.
Study the pictures. Then answer the questions.

Saul's money

Perry's money

1. How much money does each person have?

Saul has $____________ Perry has $____________

2. Saul buys gum for 75 cents. If he pays with a $1.00 bill, how much change will he get?

a. 75 cents b. $1.00 c. 25 cents d. 50 cents

3. Perry bought a banana and an apple for a total of 83 cents. How much change should he get for a $1.00 bill?

a. 27 cents b. 17 cents c. 33 cents d. 83 cents

Draw the coins Perry could get in change.

Skills: • identifies and compares fractional parts through 10ths
• identifies fractions equal to 1

Directions: Choose or write the correct answers to the questions.

1. What fraction is shaded?

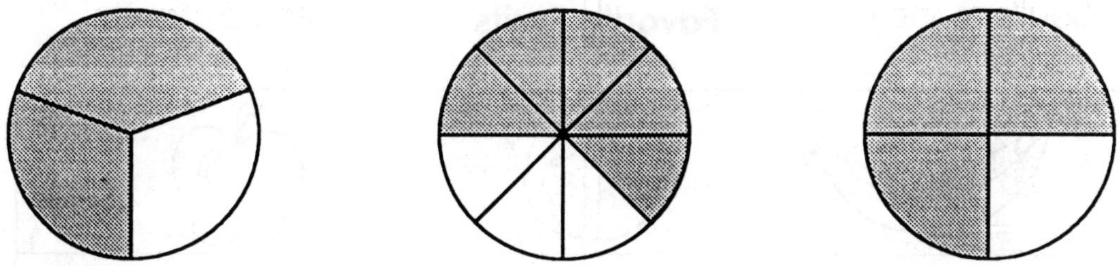

__________ __________ __________

2. Put the fractions in order from smallest to largest.

$\frac{1}{4}$ $\frac{1}{5}$ $\frac{1}{9}$ $\frac{1}{2}$ $\frac{1}{3}$ $\frac{1}{10}$

Smallest ____ ____ ____ ____ ____ ____ Largest

3. Circle the fractions that are equal to 1:

$\frac{2}{2}$ $\frac{7}{8}$ $\frac{1}{3}$ $\frac{9}{10}$ $\frac{4}{4}$ $\frac{7}{7}$

4. Use objects to help you solve these problems:

a. $\frac{1}{4}$ of 16 is ________ b. $\frac{1}{2}$ of 10 is ________

c. $\frac{1}{3}$ of 12 is ________ d. $\frac{1}{5}$ of 10 is ________

3 REVIEW

Skill: interprets information from picture and bar graphs

Directions: Write or choose the correct answers to the questions.
Study the graphs. Then answer the questions.

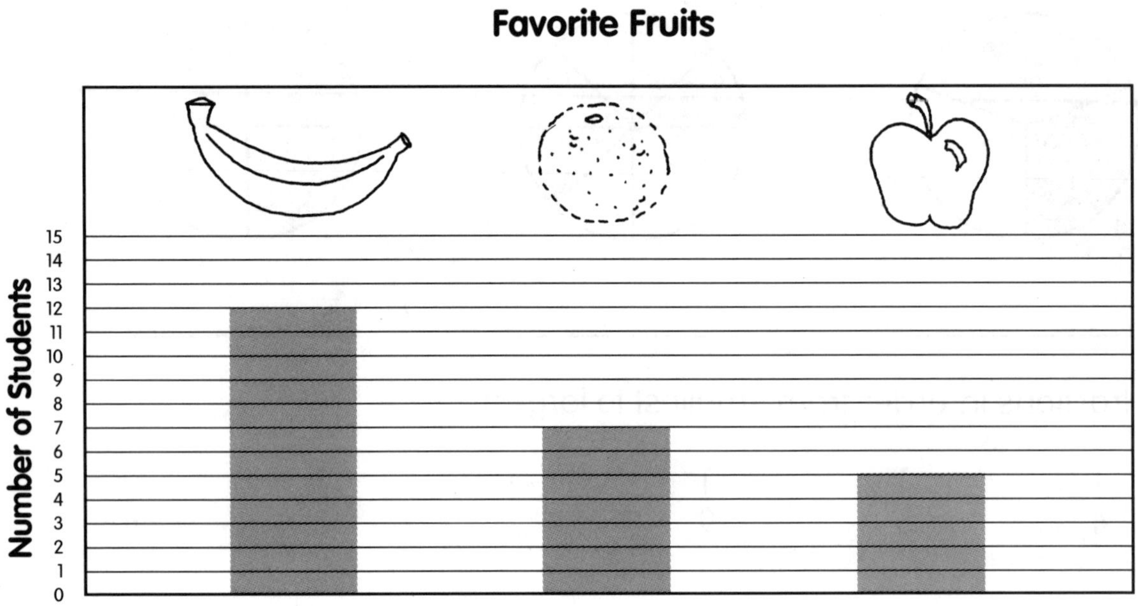

1. Which fruit did 7 students prefer? _______________

2. Which fruit is most popular? _______________

3. How many more students like bananas than like apples? _______________

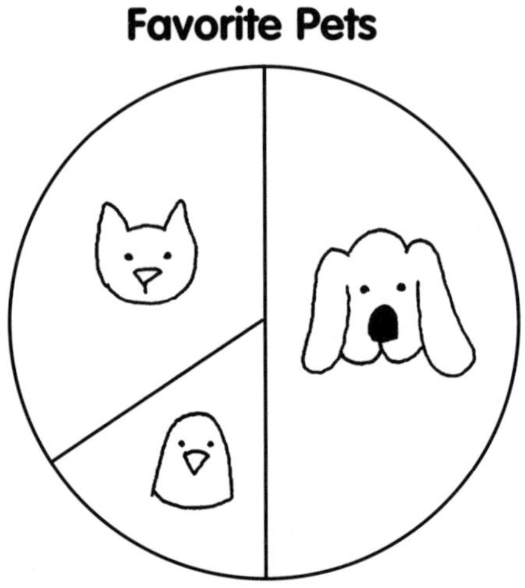

1. What pet do half the people own? _______________

2. What is the least favorite pet? _______________

Skills: knows multiplication and division facts through the nines

REVIEW 3

Directions: Do not use your calculator for this page.

1. Multiply

6 x 6 =	9 x 6 =	8 x 7 =
7 x 7 =	8 x 8 =	8 x 9 =

2. Divide

54 ÷ 6 =	48 ÷ 8 =	72 ÷ 9 =
81 ÷ 9 =	63 ÷ 7 =	36 ÷ 6 =

3. Multiply

$$\begin{array}{r} 7 \\ \underline{\times\ 6} \end{array} \qquad \begin{array}{r} 9 \\ \underline{\times\ 7} \end{array} \qquad \begin{array}{r} 9 \\ \underline{\times\ 9} \end{array} \qquad \begin{array}{r} 8 \\ \underline{\times\ 6} \end{array}$$

4. Divide

$9\overline{)63}$ $\qquad 9\overline{)54}$ $\qquad 6\overline{)48}$ $\qquad 8\overline{)72}$

3 REVIEW

Skill: uses multiplication and division facts through the 9's to solve word problems

Directions: Write and solve each problem in the space provided. Do not use a calculator. Write your answer on the line. Be sure to label your answer. For example:

Bob, Carlos, and Frank want to evenly divide 15 marbles. How many marbles will each get?

15 ÷ 3 = 5 <u>5 marbles each</u>

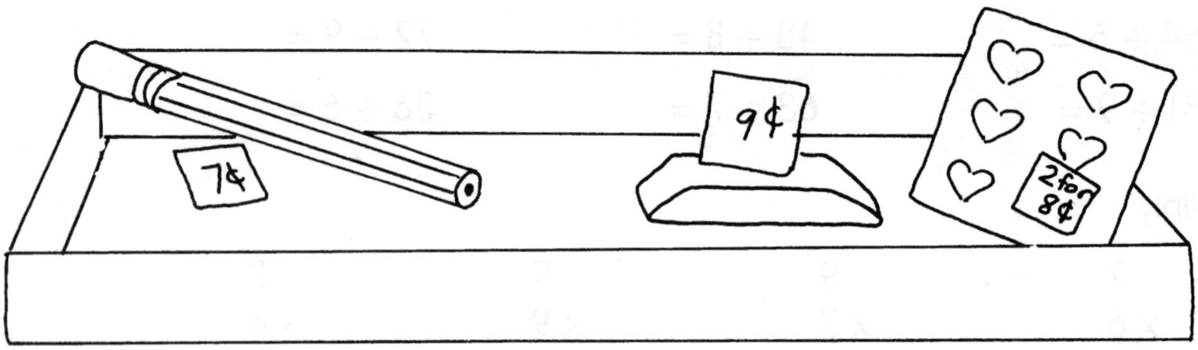

1. How much to buy 6 pencils? ____________

2. How much to buy 12 stickers? ____________

3. I have 55¢. Which of these can I buy - 7 erasers or 7 pencils? ____________

4. Angela has 64¢. How many stickers can she get? ____________

5. I want to buy 3 pencils and 2 erasers. How much money do I need? ____________

These are the coins I have. Do I have enough money? ____________

REVIEW 3

Skill: identifies information needed and not needed to solve a problem

Directions: Read each story carefully. Then answer the questions.

1. On Saturday, Lavelle went to the beach with her friend, Karina. Lavelle is 9 years old and Karina is 10. The girls left home at about 1:00 P.M. They stayed at the beach for 3 hours. At what time did they leave the beach?

Which sentence is not needed to answer the question?
a. Lavelle is 9 years old and Karina is 10.
b. The girls left home ar about 1:00 P.M.
c. They stayed at the beach for 3 hours.

2. The factory makes cars every day. The colors of the cars are black, white, red and green. Black cars are the favorite of most Americans. How many days would it take the factory to make 500 cars?

Which sentence needs to be added to answer the question?
a. More men than women drive cars.
b. White cars are easy to see at night.
c. The factory can make 100 cars every day.

3. The school choir will have a concert in the auditorium. The auditorium holds 300 chairs. How many rows of chairs will there be for the concert?

Which sentence needs to be added to answer the question?
a. 275 people will come to the concert.
b. There will be 50 chairs in each row.
c. 125 children will sing.

3 REVIEW

Skills: • multiplies a two digit number by a one digit number with no regrouping (carrying)
• solves one step problems involving multiplication

Example: $\begin{array}{r} 30 \\ \underline{\times 3} \\ 90 \end{array}$

Directions: Solve the problems below. Do not use a calculator.

1. $\begin{array}{r} 42 \\ \underline{\times 2} \end{array}$ $\begin{array}{r} 14 \\ \underline{\times 2} \end{array}$ $\begin{array}{r} 33 \\ \underline{\times 3} \end{array}$ $\begin{array}{r} 20 \\ \underline{\times 4} \end{array}$

2. $\begin{array}{r} 11 \\ \underline{\times 7} \end{array}$ $\begin{array}{r} 10 \\ \underline{\times 6} \end{array}$ $\begin{array}{r} 23 \\ \underline{\times 2} \end{array}$ $\begin{array}{r} 32 \\ \underline{\times 3} \end{array}$

3. There are 6 people in the White family. Each of them ate 11 pretzels as a snack. How many pretzels were eaten?

__________ pretzels

4. Each of the 24 kids in my class gets 2 new pencils every month. How many pencils are used each month?

__________ pencils

5. When Robert goes to the playground, he walks 4 blocks round trip. How many blocks will he walk if he goes to the playground 12 times?

__________ blocks

REVIEW 3

Skill: • multiplies a two digit number by a one digit number with regrouping (carrying)
• solves one step problems involving multiplication

Example:

$$\begin{array}{r} {}^{1}34 \\ \underline{\times\ 4} \\ 136 \end{array}$$

Directions: Work each problem or choose the right answer.

1. $\begin{array}{r} 67 \\ \underline{\times\ 5} \end{array}$ $\begin{array}{r} 43 \\ \underline{\times\ 8} \end{array}$ $\begin{array}{r} 19 \\ \underline{\times\ 7} \end{array}$ $\begin{array}{r} 36 \\ \underline{\times\ 3} \end{array}$ $\begin{array}{r} 45 \\ \underline{\times\ 5} \end{array}$

2. $\begin{array}{r} 28 \\ \underline{\times\ 7} \end{array}$ $\begin{array}{r} 14 \\ \underline{\times\ 5} \end{array}$ $\begin{array}{r} 29 \\ \underline{\times\ 3} \end{array}$ $\begin{array}{r} 17 \\ \underline{\times\ 7} \end{array}$ $\begin{array}{r} 34 \\ \underline{\times\ 3} \end{array}$

3. Meadowview School has 5 classrooms at each grade. Each classroom has 28 students. How many first graders will go to Meadowview School?

a. 33 b. 28 c. 100 d. 140

4. Four school buses bring children to Meadowview School. Each bus can carry 56 children. How many children can ride the bus to school?

a. 114 b. 224 c. 60 d. 204

3 REVIEW

Parents: The skills on this page are challenging and may not have been taught in your child's third grade class.

Skills: • divides a two-digit number by a one-digit number with and without remainders
• solve two-step word problems

Directions: Solve the problems below. Do not use a calculator

1.	$8\overline{)30}$	$7\overline{)44}$	$5\overline{)26}$	$9\overline{)59}$
2.	$4\overline{)56}$	$6\overline{)90}$	$3\overline{)72}$	$5\overline{)65}$

Directions: Solve these word problems. You may use a calculator to help you.

1. You have decided to buy a new video game that is on sale for $39.95. Your bank contains $17.75. If you can earn $4.50 a week walking the neighbors dog, how many weeks will it be before you can buy the video?

work space:

______________weeks

2. Super Save Food Market is selling ice cream for $2.49 a quart. Best Buy Groceries is selling ice cream for $1.69 a pint. Where will you go for the best buy? How much will you save if you buy a half gallon?

work space:

I will go to ________________________.

I will save ______________ on a half gallon.

Answer Key

Please take time to go over the work your child has completed. Ask your child to explain what he/she has done. Praise both success and effort. If mistakes have been made, explain what the answer should have been and how to find it. Let your child know that mistakes are a part of learning. The time you spend with your child helps let him/her know you feel learning is important.

page 1

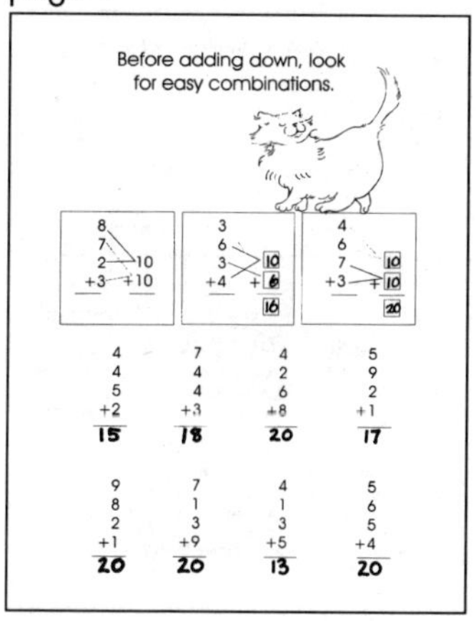

Before adding down, look for easy combinations.

8, 7, 2, +3 → 10, 10
3, 6, 3, +4 → 10, 6, 16
4, 6, 7, +3 → 10, 10, 20

4 + 4 + 5 + 2 = 15
7 + 4 + 4 + 3 = 18
4 + 2 + 6 + 8 = 20
5 + 9 + 2 + 1 = 17

9 + 8 + 2 + 1 = 20
7 + 1 + 3 + 9 = 20
4 + 1 + 3 + 5 = 13
5 + 6 + 5 + 4 = 20

page 2

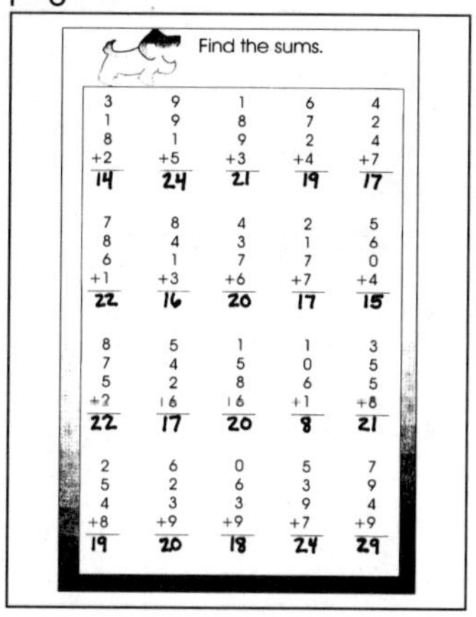

Find the sums.

3 + 1 + 8 + 2 = 14
9 + 9 + 1 + 5 = 24
1 + 8 + 9 + 3 = 21
6 + 7 + 2 + 4 = 19
4 + 2 + 4 + 7 = 17

7 + 8 + 6 + 1 = 22
8 + 4 + 1 + 3 = 16
4 + 3 + 7 + 6 = 20
2 + 1 + 7 + 7 = 17
5 + 6 + 0 + 4 = 15

8 + 7 + 5 + 2 = 22
5 + 4 + 2 + 6 = 17
1 + 5 + 8 + 6 = 20
1 + 0 + 6 + 1 = 8
3 + 5 + 5 + 8 = 21

2 + 5 + 4 + 8 = 19
6 + 2 + 3 + 9 = 20
0 + 6 + 3 + 9 = 18
5 + 3 + 9 + 7 = 24
7 + 9 + 4 + 9 = 29

page 3

Add down, then up to check your answers.

1 + 2 + 3 + 4
7 + 4 + 9 + 1 = 21, 21
6 + 5 + 1 + 2 = 14, 14
3 + 7 + 1 + 8 = 19, 19
9 + 6 + 8 + 3 = 26, 26
4 + 8 + 3 + 2 = 17, 17
7 + 5 + 6 + 4 = 22, 22
5 + 9 + 4 + 6 = 24, 24
4 + 7 + 2 + 3 = 16, 16

page 4

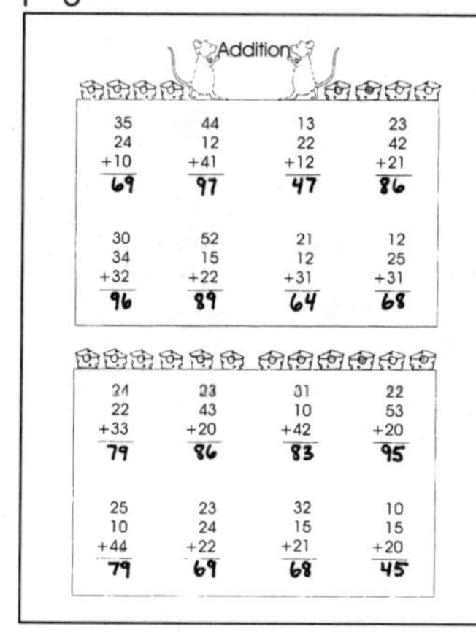

Addition

35 + 24 + 10 = 69
44 + 12 + 41 = 97
13 + 22 + 12 = 47
23 + 42 + 21 = 86

30 + 34 + 32 = 96
52 + 15 + 22 = 89
21 + 12 + 31 = 64
12 + 25 + 31 = 68

24 + 22 + 33 = 79
23 + 43 + 20 = 86
31 + 10 + 42 = 83
22 + 53 + 20 = 95

25 + 10 + 44 = 79
23 + 24 + 22 = 69
32 + 15 + 21 = 68
10 + 15 + 20 = 45

page 5

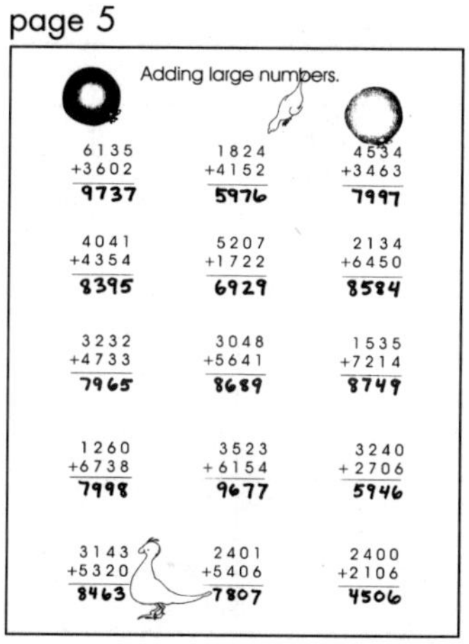

Adding large numbers.

6135 + 3602 = 9737
1824 + 4152 = 5976
4534 + 3463 = 7997

4041 + 4354 = 8395
5207 + 1722 = 6929
2134 + 6450 = 8584

3232 + 4733 = 7965
3048 + 5641 = 8689
1535 + 7214 = 8749

1260 + 6738 = 7998
3523 + 6154 = 9677
3240 + 2706 = 5946

3143 + 5320 = 8463
2401 + 5406 = 7807
2400 + 2106 = 4506

page 6

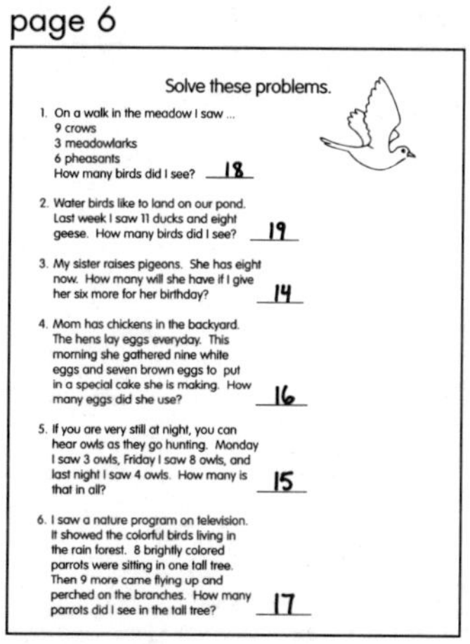

Solve these problems.

1. On a walk in the meadow I saw ... 9 crows, 3 meadowlarks, 6 pheasants. How many birds did I see? 18
2. Water birds like to land on our pond. Last week I saw 11 ducks and eight geese. How many birds did I see? 19
3. My sister raises pigeons. She has eight now. How many will she have if I give her six more for her birthday? 14
4. Mom has chickens in the backyard. The hens lay eggs everyday. This morning she gathered nine white eggs and seven brown eggs to put in a special cake she is making. How many eggs did she use? 16
5. If you are very still at night, you can hear owls as they go hunting. Monday I saw 3 owls, Friday I saw 8 owls, and last night I saw 4 owls. How many is that in all? 15
6. I saw a nature program on television. It showed the colorful birds living in the rain forest. 8 brightly colored parrots were sitting in one tall tree. Then 9 more came flying up and perched on the branches. How many parrots did I see in the tall tree? 17

page 7

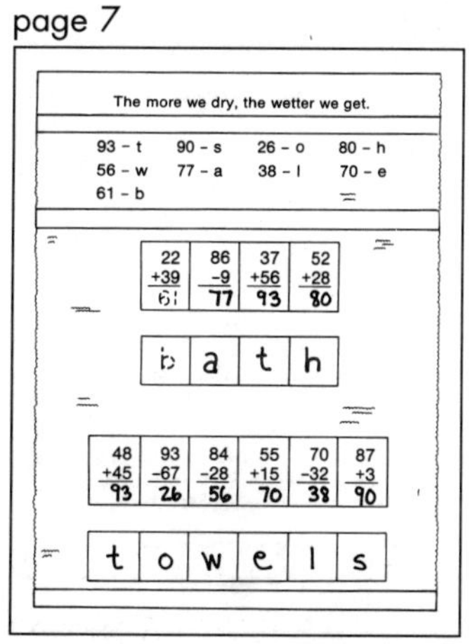

The more we dry, the wetter we get.

93 - t, 90 - s, 26 - o, 80 - h
56 - w, 77 - a, 38 - l, 70 - e
61 - b

22 + 39	86 − 9	37 + 56	52 + 28
61	77	93	80
b	a	t	h

48 + 45	93 − 67	84 − 28	55 + 15	70 − 32	87 + 3
93	26	56	70	38	90
t	o	w	e	l	s

page 8

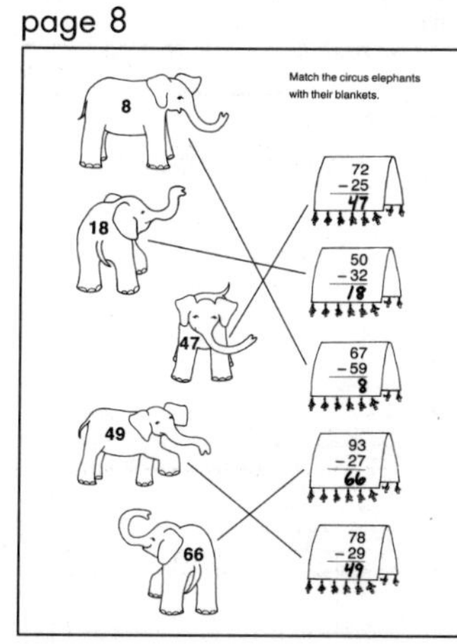

Match the circus elephants with their blankets.

Elephants: 8, 18, 47, 49, 66
Blankets: 72 − 25 = 47; 50 − 32 = 18; 67 − 59 = 8; 93 − 27 = 66; 78 − 29 = 49

page 9

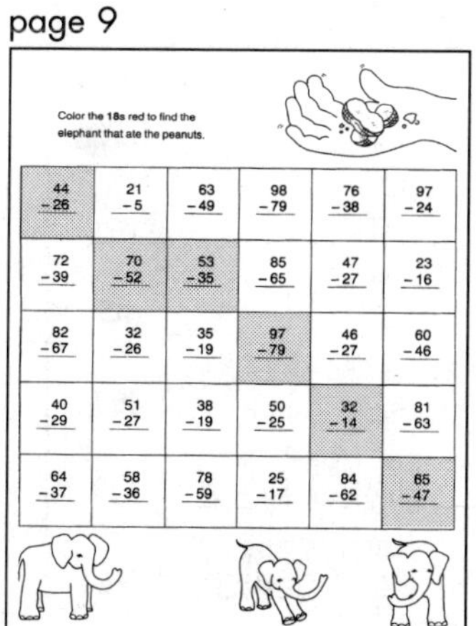

Color the 18s red to find the elephant that ate the peanuts.

44 − 26	21 − 5	63 − 49	98 − 79	76 − 38	97 − 24
72 − 39	70 − 52	53 − 35	85 − 65	47 − 27	23 − 16
82 − 67	32 − 26	35 − 19	97 − 79	46 − 27	60 − 46
40 − 29	51 − 27	38 − 19	50 − 25	32 − 14	81 − 63
64 − 37	58 − 36	78 − 59	25 − 17	84 − 62	65 − 47

page 10

Elephants for Sale

$41.04, $12.25, $6.10, $14.30, $4.65, $22.22, $5.23

How much will we cost?

$4.65 + 6.10 = $10.75
$5.23 + 4.65 = $9.88
$12.25 + 22.22 = $34.47
$41.04 + 14.30 = $55.34

$41.04 + 22.22 + 5.23 = $68.49
$4.65 + 5.23 + 6.10 = $15.98
$12.25 + 14.30 + 41.04 = $67.59
$22.22 + 13.30 + 12.25 = $47.77

How much more do we cost?

$8.69, $37.58

$8.69 − 6.10 = $2.59
$8.69 − 4.69 = $4.00
$37.58 − 22.22 = $15.36
$37.58 − 14.30 = $23.28

page 11

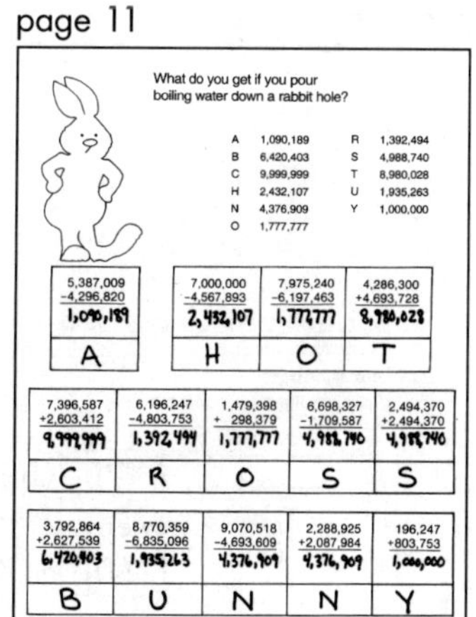

What do you get if you pour boiling water down a rabbit hole?

A 1,090,189; B 6,420,403; C 9,999,999; H 2,432,107; N 4,376,909; O 1,777,777; R 1,392,494; S 4,988,740; T 8,980,028; U 1,935,263; Y 1,000,000

5,387,009 − 4,296,820	7,000,000 − 4,567,893	7,975,240 − 6,197,463	4,286,300 + 4,693,728
1,090,189	2,432,107	1,777,777	8,980,028
A	H	O	T

7,396,587 + 2,603,412	6,196,247 − 4,803,753	1,479,398 + 298,379	6,698,327 − 1,709,587	2,494,370 + 2,494,370
9,999,999	1,392,494	1,777,777	4,988,740	4,988,740
C	R	O	S	S

3,792,864 + 2,627,539	8,770,359 − 6,835,096	9,070,518 − 4,693,609	2,288,925 + 2,087,984	196,247 + 803,753
6,420,403	1,935,263	4,376,909	4,376,909	1,000,000
B	U	N	N	Y

page 12

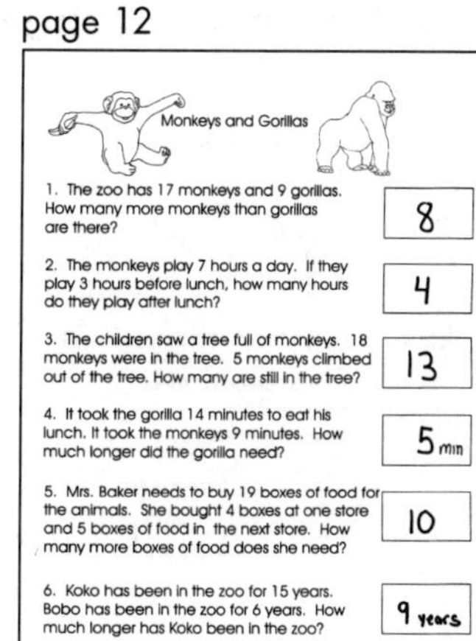

Monkeys and Gorillas

1. The zoo has 17 monkeys and 9 gorillas. How many more monkeys than gorillas are there? 8
2. The monkeys play 7 hours a day. If they play 3 hours before lunch, how many hours do they play after lunch? 4
3. The children saw a tree full of monkeys. 18 monkeys were in the tree. 5 monkeys climbed out of the tree. How many are still in the tree? 13
4. It took the gorilla 14 minutes to eat his lunch. It took the monkeys 9 minutes. How much longer did the gorilla need? 5 min
5. Mrs. Baker needs to buy 19 boxes of food for the animals. She bought 4 boxes at one store and 5 boxes of food in the next store. How many more boxes of food does she need? 10
6. Koko has been in the zoo for 15 years. Bobo has been in the zoo for 6 years. How much longer has Koko been in the zoo? 9 years

page 13

Find the mistakes and correct them.

68 - 12 = 56	45 - 30 = 15	28 - 24 = ~~4~~ 4	34 - 11 = 23
27 - 10 = 17	42 - 22 = ~~~~ 20	68 - 37 = ~~~~ 31	45 - 13 = 32
79 - 45 = ~~~~ 34	66 - 32 = 34	58 - 50 = 8	73 - 41 = ~~~~ 32
86 - 55 = 31	65 - 14 = 51	79 - 46 = ~~~~ 33	59 - 37 = ~~~~ 22

page 14

Subtract

837 - 614	398 - 235 = 163	254 - 122 = 132	473 - 351 = 122
256 - 142 = 114	638 - 231 = 407	356 - 214 = 142	189 - 123 = 66
643 - 331 = 312	532 - 210 = 322	975 - 134 = 841	837 - 213 = 624
584 - 121 = 463	324 - 102 = 222	949 - 325 = 624	613 - 312 = 301

page 15

How Many?

Use your ruler and measuring tape to help you answer these questions.

Match:

- inches in a yard — 36
- centimeters in a meter — 100
- feet in a yard — 3
- inches in a foot — 12

Answer these word problems.

1. How many inches in 2 feet? 24
2. How many feet in 5 yards? 15
3. How many centimeters in 3 meters? 300
4. How many inches in 2 yards? 72

page 16

Parents: Your child will need a set of measuring spoons and cups to do this page. Have paper towels handy since water may be spilled.

Measuring Cups and Spoons

Use water and measuring cups and spoons to help you figure out the answers to these questions.

Measuring Cups: 4, 3, 2, 6, 3, 6

Measuring Spoons: 3, 2, 6, 4, 12, 1

page 17

Riddles

Paste the answer under each riddle.

I am used to measure large things like lumber.	I am used to measure flour when you bake.
I am used to measure pieces of cloth.	I am used when you want to draw a straight line.
I am a container for milk and juice.	I am used to measure little things like salt and spices.

page 18

Parents: Encourage your child to draw out the problems to find the answers.

Find the Answers

Five people are going on a beach picnic.

Answer these problems to find out food and drink will they need.

1. There are 3 glasses of water in each liter bottle. How many liters do you need if 3 people drink 2 glasses of water and 1 person drinks 3 glasses of water? 3 liters
2. It takes 2 tablespoons of peanut butter and 3 tablespoons of jelly to make a sandwich. How much peanut butter do you need for 5 sandwiches? 10 tablespoons How much jelly do you need for 5 sandwiches? 15 tablespoons
3. Each person will drink two cups of lemonade. How many cups do you need? 10 cups

page 19

How Heavy Is It?

16 ounces = 1 pound

Some things you buy are measured by weight.

- packages at the post office
- dog food at the pet store
- nails at some hardware stores
- food at the supermarket

In the United States these are measured in **pounds** and **ounces**. In most other countries they are measured in **grams**.

Look around your house and garage.

1. Find something that is measured in: pounds, ounces, grams
2. Find something that weighs about: 5 pounds, 8 ounces, 1 pound, 16 ounces, 12 ounces, 227 grams

answers will vary

page 20

When do you go to bed?

Look at this graph. Answer these questions.

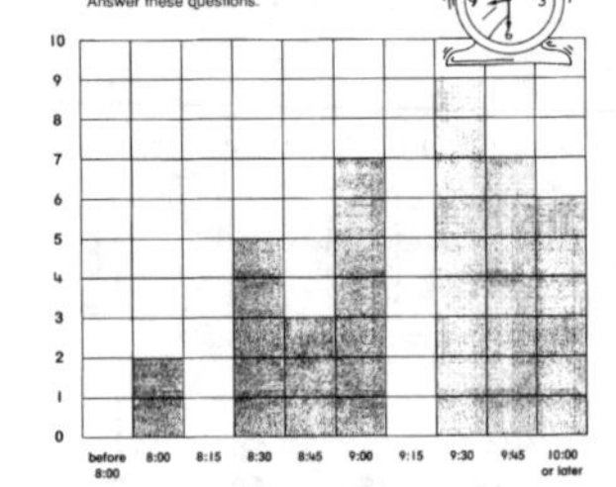

1. What is the earliest anyone went to bed? 8:00
2. How many people went to bed at 10:00 or later? 6
3. At what time did the most people go to bed? 9:30
4. What time do you go to bed? Answers will vary Show the time on this graph. Mark your box like this:
5. At what times did no one go to bed? before 8:00, 8:15, 9:15

page 21

Make a Bar Graph

Collect the information listed below and complete this bar graph.

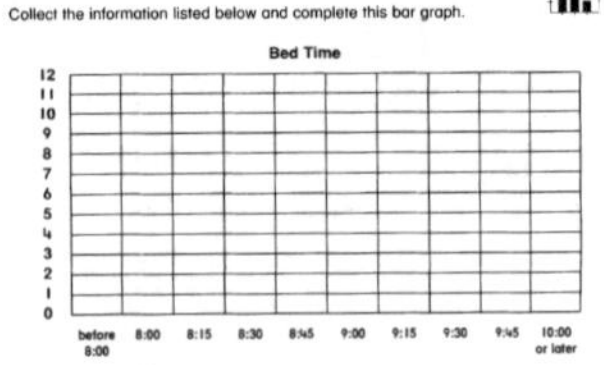

1. Collecting information and making the graph.
   a. Color a box on the graph to show your bed time.
   b. Ask each person in your family, "What time did you go to bed last night?" Color one box on the graph for each answer.
   c. Ask 10 more people you know the same question. Color one box on the graph for each answer.

   answers will vary
2. Answer questions about your graph.
   a. What time did the most people go to bed?
   b. How many people went to bed at the same time as you?
   c. What was the latest time anyone you asked goes to bed?
   d. What was the earliest time anyone you asked goes to bed?

page 22

Parts of a Bicycle

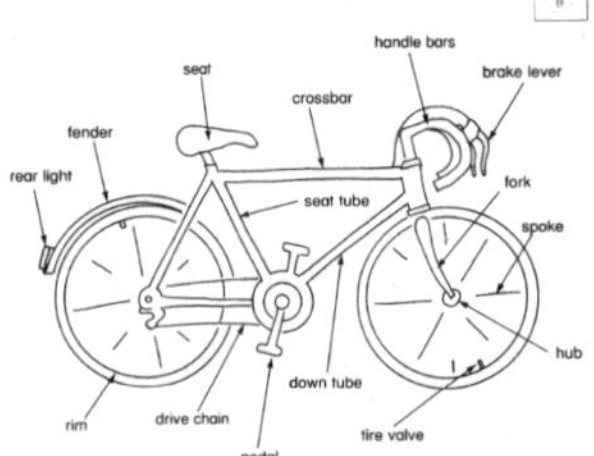

1. What do you call the part of the bicycle that holds the seat? Seat tube
2. Where do you put the air in a bicycle tire? tire valve
3. Name three parts of the bicycle wheel. Spoke, hub, rim
4. What is the part of a bicycle where you put your feet called? pedal
5. The rear light is on the fender

page 23

Big Sale!

Read the graph to help you answer the questions below.

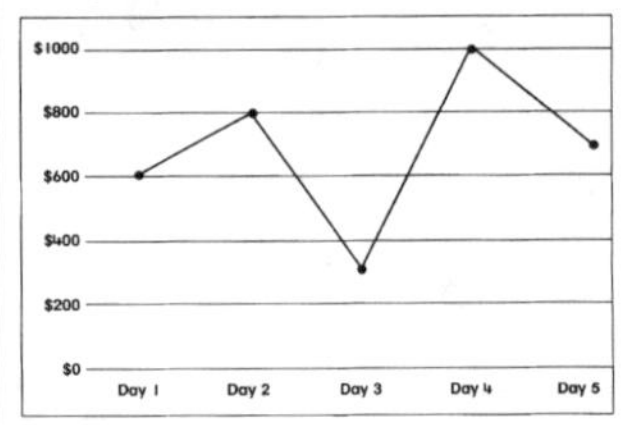

1. How much was sold on ______?
   Day 3 $300
   Day 1 $600
   Day 5 $700
2. On what day was the greatest amount sold? Day 4
3. On what day was the smallest amount sold? Day 3
4. How much more was sold on Day 4 than on Day 1? $400

Bonus

5. How much was sold all together? $3,400

page 24

Make a Circle Graph

I asked 24 children "How do you get to school?" I collected these answers:

walk—12, ride bike—3, ride bus—6, ride in a car—3

Record this information on the circle graph below. Color in the part of the story that shows the number for each way the children came to school.

walk—red, bike—blue, bus—yellow, car—green

Hint: Each piece of the circle stands for 3 children.

page 25

I can see 15 people on the street.
5 have open umbrellas.
7 have closed umbrellas.

How many don't have an umbrella?

I can solve the problem. 3 People

page 26

Look at the pattern.
How many flowers will go in the next vase?

I can solve the problem. 8 flowers

page 27

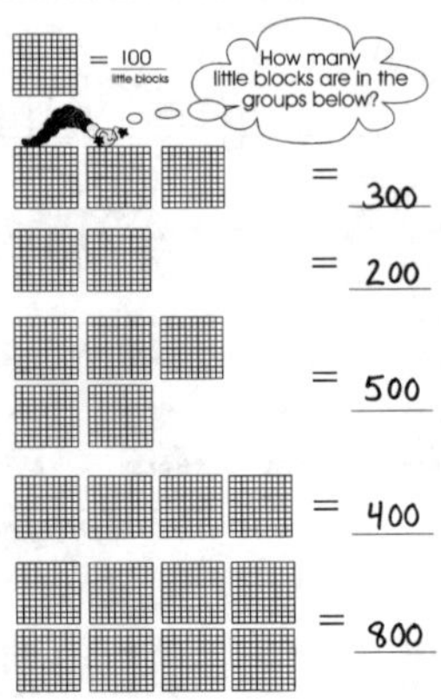

= 100 little blocks

How many little blocks are in the groups below?

= 300
= 200
= 500
= 400
= 800

page 28

More than 100.

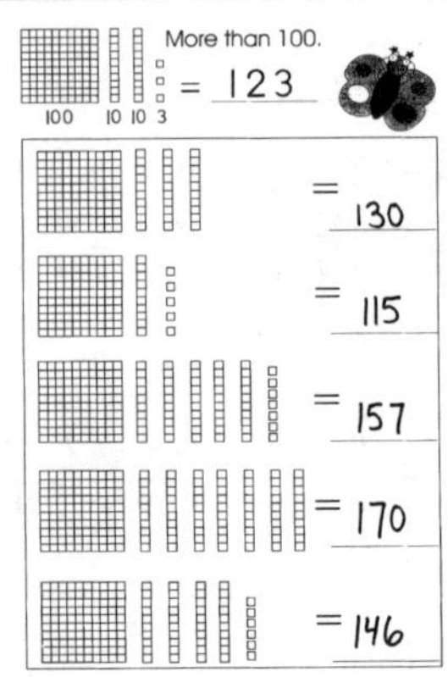

100 10 10 3 = 123

= 130
= 115
= 157
= 170
= 146

page 29

Counting BIG Amounts

= 234

= 341

= 254

= 402

= 300

= 232

= 410

page 30

Think about what you know about hundreds, tens, and ones.

5 hundreds + 6 tens + 3 ones = 563

6 hundreds + 8 tens + 6 ones = 686

7 hundreds + 4 tens + 9 ones = 749

8 hundreds + 2 tens + 2 ones = 822

9 hundreds + 6 tens + 4 ones = 964

8 hundreds + 1 tens + 8 ones = 818

7 hundreds + 7 tens + 1 ones = 771

6 hundreds + 9 tens + 5 ones = 695

5 hundreds + 3 tens + 7 ones = 537

page 31

Skills: • identifies numbers both in word and numeral form
• identifies odd and even numbers

Examples: these numbers are odd - 1, 3, 17, 25
these numbers are even - 4, 8, 16, 26

Directions: Circle or write the correct answer for each question.

1. Which number is three hundred twenty-seven? a. 723 (b. 327) c. 3270 d. 237
2. Which number is nine thousand forty? a. 904 b. 9004 c. 940 (d. 9040)
3. Which number is eighty-four thousand two hundred? a. 80,200 (b. 84,200) c. 8420 (d. 84,200)
4. Draw lines to match each number with its word name: 630 – six hundred thirty; 6300 – six thousand three hundred; 63 – sixty-three; 603 – six hundred three
5. Circle the odd numbers: (95) 34 28 (43) (81) 18
6. Circle the even numbers: (98) (82) (26) 35 39 (64)

page 32

Skills: • identifies ordinal numbers
• understands expanded form
• uses (greater than) (less than) symbols correctly

Examples: ordinal numbers - first, sixth etc.
expanded form - 467 = 400 + 60 + 7
63 > 60, 60 < 63

Directions: Choose or write the correct answers to each question.

1. Rewrite in the correct order: Fourth, Sixth, First, Second, Third, Fifth — First, Second, Third, Fourth, Fifth, Sixth
2. Rewrite each number in expanded form: For example: 562 = 500+60+2
   78 — 70 + 8
   245 — 200 + 40 + 5
3. Read the expanded form; then write the number. For example: 200+30+7 = 237
   800 + 20 + 6 — 826
   100 + 90 + 9 — 199
4. Put > or < in each circle:
   59 > 61 (answer shown: 59 < 61); 304 < 314; 145 > 135; 5972 > 5952

page 33

Skills: • shows understanding of number order
• can repeat a number pattern
• can use +, -, x signs to make true number statements

Directions: Write the correct answers to each question.

1. Rewrite the numbers in the correct order from smallest to largest: 839 29 391 48 85 17 8 — 8 17 29 48 85 391 839 (smallest … largest)
2. Continue the pattern:
   4 7 10 13 16
   55 53 51 49 47 45 43
   3, 4, 5 13, 14, 15 23, 24, 25 33, 34, 35
3. Write the number that is one before and one after the numbers shown:
   419 420 421; 398 399 400
   1049 1050 1051; 7356 7357 7358
   94,670 94,671 94,672; 26,518 26,519 26,520
4. Put +, -, or x in each circle to make the number statement true:
   3 x 5 = 15; 17 - 9 = 8
   6 + 4 + 8 = 18; 6 + 4 - 8 = 2

page 34

Skills: • counts by 2s, 5s and 10s to 100
• understands place value

Directions: Write the correct answer to each question.

1. Write the numbers to count by 2s: 22 24 26 28 30 32 34
2. Write the numbers to count by 5s: 50 55 60 65 70 75 80
3. Write the numbers to count by 10s: 40 50 60 70 80 90 100
4. Round each number to the nearest 10: 58 = almost 60; 61 = almost 60; 29 = almost 30
5. Look carefully at each number. Circle the number in the ones place; put an X on the number in the hundreds place.
6. Look carefully at each number. Put a box around the number in the tens place; put a line under the number in the thousands place.
7. Write the numbers for each of these:
   4 tens and 8 ones = 48
   4 hundreds, 5 tens and 2 ones = 452
   8 thousands, 7 hundreds, 0 tens and 6 ones = 8706
   4 thousands, 0 hundreds, 3 tens = 4030

page 35

Skills: • adds and subtracts single digit numbers in a series
• adds and subtracts multi-digit numbers without borrowing and carrying

Directions: Do not use your calculator for this page.

1. Add or subtract. Watch the signs!
   3 + 4 + 7 = 14; 5 - 2 + 8 = 11
   3 + 5 - 7 + 9 - 2 = 8; 19 - 3 + 5 - 1 + 4 = 24
2. 49 - 21 = 28; 30 + 22 = 52; 384 - 103 = 281; 994 - 4 = 990; 4651 + 1326 = 5977
3. 845 - 31 = 814; 75 + 22 = 97; 829 - 726 = 103; 104 + 14 = 118; 9805 - 2300 = 7505
4. Study the picture. Then answer the questions.

How much will you need to buy the ball and the truck? 89¢

How much more does the doll cost than the yo-yo? 74¢ more

page 36

Skill: solves one-step word problems requiring addition and subtraction without regrouping (carrying and borrowing)

Directions: Read the information. Then work each problem in the space provided. Write your answer on the line. Be sure to label your answer, for example: 8 big dogs. 4 small dogs. How many dogs? 8 + 4 = 12 12 dogs

The four third grade classes are going on a trip to a park. Two classes have 23 children and two classes have 21 children.

1. How many children will go on the trip to the park? 88 children
2. A school bus holds 50 children. How many buses will be needed? 2 buses
3. It takes 24 minutes to go from the school to the park. How long is a round trip? 48 minutes
4. At the park 24 children want to go for a hike. How many will stay behind? 64 children
5. The park has 7 swings and a merry-go-round that holds 12 children. Can Ms. Jones class of 21 students use the equipment at the same time? NO

page 37

Skill: adds and subtracts multi-digit numbers with regrouping (carrying and borrowing)

Directions: Add or subtract. Do not use your calculator. Watch the signs!

1. 57 + 35 = 92; 86 + 47 = 133; 325 - 69 = 256; 700 - 450 = 250; 588 + 281 = 869
2. 836 - 379 = 457; 5066 + 1789 = 6855; 4290 + 3876 = 8166; 9429 - 5683 = 3746; 600 - 151 = 449
3. Study the picture. Then answer the questions.

At the Movies

Children under 12	$2.75
Adults	$6.50
Senior Citizen (62 & up)	$4.25
Silver Screen Passes	$5.00

How much for 2 adults to see a movie? $13.00

How much for Grandma (age 67) to take Sarah (age 9)? $7.00

page 38

Skill: solves one-step word problems requiring addition and subtraction with regrouping (carrying and borrowing)

Directions: Use the information on the map to work each problem. Do not use a calculator. Write and work each problem in the space provided. Write your answer on the line. Be sure to label your answer, for example: 8 big dogs. 4 small dogs. How many dogs? 8 + 4 = 12 12 dogs

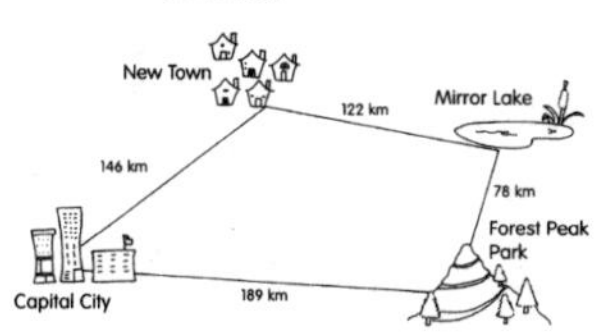

1. Mr. Diaz delivers packages. Every day he drives from Capital City to New Town. Then he drives on to Mirror Lake, Forest Peak Park and finally back to Capital City. How far does he drive every day? 535 Km
2. Jose and Tina live in Capital City. They want to drive to Mirror Lake. Which way is shorter - through New Town or through Forest Peak Park? Forest Peak Park
3. Ms. Lee has enough gas in her car to drive 250 km. Can she get from New Town to Forest Peak Park without stopping to refuel? Yes

page 39

Skills: knows multiplication and division facts through the 5's

Directions: Do not use your calculator for this page.

1. Multiply
   4 x 3 = 12; 5 x 2 = 10; 6 x 4 = 24
   7 x 5 = 35; 4 x 4 = 16; 2 x 6 = 12
2. Divide
   30 ÷ 5 = 6; 18 ÷ 2 = 9; 24 ÷ 4 = 6
   15 ÷ 3 = 5; 20 ÷ 5 = 4; 9 ÷ 3 = 3
3. Multiply
   6 x 3 = 18; 5 x 5 = 25; 9 x 4 = 36; 8 x 3 = 24; 3 x 3 = 9
4. Divide
   14 ÷ 2 = 7; 18 ÷ 3 = 6; 12 ÷ 4 = 3; 45 ÷ 5 = 9; 36 ÷ 4 = 9; 21 ÷ 3 = 7

page 40

Skills: • understands basic properties of multiplication
• understands the relationship between multiplication and division

Directions: Solve the problems without using a calculator.

1. 2 x 3 = 3 x 2; 6 x 7 = 7 x 6
2. (3 x 4) x 2 = 24; 3 x (4 x 2) = 24
3. 2 x 1 = 2; 5 x 1 = 5; 1 x 6 = 6; Any number multiplied by 1 = that number
4. 4 x 0 = 0; 9 x 0 = 0; 0 x 5 = 0; Any number multiplied by 0 = 0
5. 9 x 2 = 18; 6 x 5 = 30 (answers will vary)
   18 ÷ 2 = 9; 18 ÷ 9 = 2; 30 ÷ 5 = 6; 30 ÷ 6 = 5

page 41

Skill: understands whether to use multiplication or division to solve a problem

Directions: Circle the operation you would use to solve each problem.

1. Samantha, Carol and Andy have 4 pieces of gum each. How much gum do they have altogether? (a. multiplication) b. division
2. There are 6 cards with 5 pictures on each. How many pictures are there? (a. multiplication) b. division
3. Each package contains 4 chocolate candies. How many packages are needed to give one candy to each of 28 students in the class? a. multiplication (b. division)
4. I have 45 photos to put in my album. I can fit 5 photos on a page. How many pages will I use up? a. multiplication (b. division)
5. Mr. Williams buys stickers that come 6 on a sheet. He has 3 sheets. How many stickers can he give out? (a. multiplication) b. division

page 42

Skill: uses multiplication and division facts through the 5's to solve word problems

Directions: Write and solve each problem in the space provided. Do not use a calculator. Write your answer on the line. Be sure to label your answer.

For example: Bob, Carlos, and Frank each have 5 marbles. How many marbles do they have together? 3 x 5 = 15 15 marbles

1. Bunches of bananas have 4, 5, or 6 bananas. Todd bought 3 bunches. What is the smallest number of bananas he might have? 12 bananas
2. What is the largest number of bananas Todd might have? 18 bananas
3. Todd just loves bananas. He eats three bananas every day. How long will 18 bananas last him? 6 days
4. Bananas are on sale for 9¢ a pound. How much will 3 pounds of bananas cost? 27¢
5. Todd has 45¢. How many pounds of bananas at 9¢ a pound can he buy? 5 pounds

page 43

Skill: uses multiplication and division facts through the 5s to solve problems

Directions: Do not use a calculator for this page.

Complete the tables: How much will it cost to buy?

	2 cents each	3 cents each	4 cents each	5 cents each
1 costs	2 cents	3 cents	4 cents	5 cents
2 costs	4 cents	6 cents	8 cents	10 cents
3 costs	6 cents	9 cents	12 cents	15 cents
4 costs	8 cents	12 cents	16 cents	20 cents
5 costs	10 cents	15 cents	20 cents	25 cents

How many French fries will each person get?

There are 25 French fries	(5 faces)	5 French fries each
There are 24 French fries	(4 faces)	6 French fries each
There are 18 French fries	(3 faces)	6 French fries each
There are 16 French fries	(2 faces)	8 French fries each

page 44

Skills: • identifies plane and space (three-dimensional) figures, points, lines, and right angles
• draws lines, points and triangles

Directions: Draw a line to match each figure with its name.

Triangle, Cube, Rectangular prism, Circle, Sphere, Square

Directions: Choose or write the correct answer to the questions.

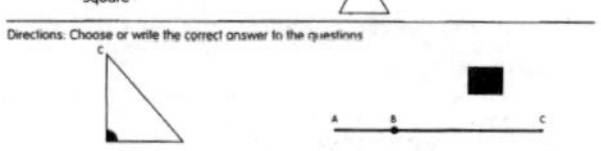

Directions: Look at Figure 1. Answer questions 1 & 2.

1. The shaded area A is a a. corner b. side (c. right angle)
2. A triangle ____ has 3 sides. a. sometimes b. never (c. always)

Look at Figure 2. Answer questions 3 - 5.

3. AC is a (a. line) b. line segment c. point
4. AB is a a. line (b. line segment) c. point
5. B is a a. line b. line segment (c. point)

page 45

Skills: recognizes congruent and symmetrical figures

Examples: Figure 1 and 2 are congruent
Figure 3 and 4 are not congruent
Figure 5 is divided symmetrically
Figure 6 is not divided symmetrically

figure 1 figure 2 figure 3 figure 4 figure 5 figure 6

Directions: Mark or show the correct answer.

1. Are the pairs of figures congruent?
   Yes (No) (Yes) No
2. Draw a figure that is congruent to the figure shown.
3. Is each figure divided symmetrically?
   (Yes) No Yes (No) Yes (No)
4. Draw a line of symmetry on this figure.

page 46

Skill: finds a location on a grid by using ordered pairs

Example: On the grid below, the ordered pair (4,6) describes the location of point A

Directions: Use the grid to answer the questions.

1. What is the ordered pair located at "B"? (2,2)
2. What is the ordered pair located at "C"? (3,1)
3. What is the ordered pair located at "E"? (7,2)
4. Put a dot at (6,1). Label it D.
5. Put a dot at (4,2). Label it F.
6. Put a dot at (6,3). Label it G.
7. Connect these points: BC CD DE BFE FA AG
8. What did you draw? a sailboat

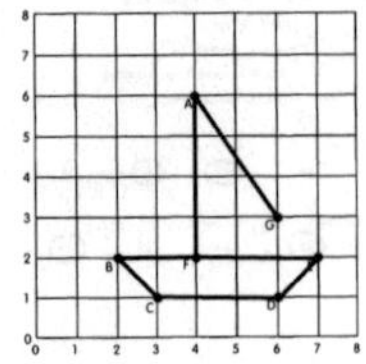

page 47

Skills:
- determines the area of a figure by counting units
- determines the perimeter of a figure

Directions: Choose or write the correct answers to the questions. Study the figures. Then answer the questions.

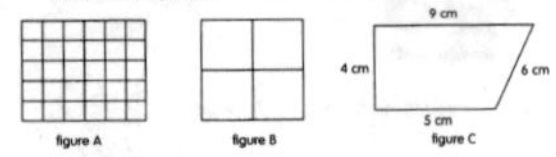

1. What is the area of Figure A?
   (a. 30 square units) b. 60 square units c. 36 square units
2. What is the area of Figure B?
   a. 8 square units (b. 4 square units) c. 16 square units
3. What is the perimeter of Figure C? 24 cm

page 48

Skills: Chooses an appropriate unit to measure for a real life situation.

Directions: What unit of measurement would you use? Choose the most sensible one.

1. To tell how far it is around the earth
   a. feet b. inches (c. miles) d. ounces
2. To tell how tall the Statue of Liberty is
   (a. meters) b. centimeters c. grams d. kilometers
3. To measure the water in a swimming pool
   a. ounces b. cups c. pints (d. gallons)
4. To tell how much juice to buy for the class party
   a. milliliters (b. liters) c. kilograms d. grams
5. To tell how much a cow weighs
   a. grams (b. kilograms) c. meters d. kilometers
6. To tell how much an apple weighs
   (a. ounces) b. pounds c. tons d. quarts

page 49

Skills:
- measures to the nearest inch or centimeter
- estimates the weight and capacity of pictured objects

Directions: Choose or write the correct answers to the questions.

Note: You will need a ruler that measures in inches and centimeters

line A line B

1. Measure line A in centimeters. 5 cm
2. Measure line B in inches. 3 inches
3. Draw a line from each picture to two ways of measuring it.

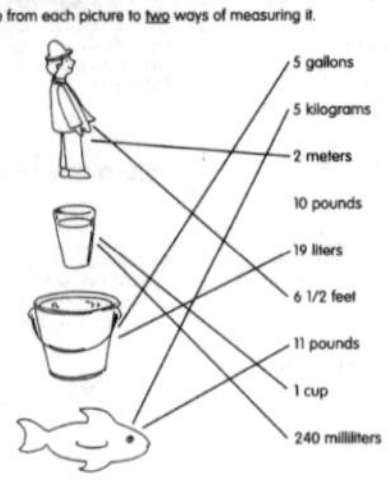

page 50

Parents: Today digital clocks are everywhere. Many children do not know how to tell time on an analog clock (a clock with hands). It's important to have analog clocks at home and to make sure your child's first watch is analog, not digital.

Skills:
- recognizes time in both analog and digital form
- tells time to the minute in more than one way
- can add or subtract from the time represented on a clock

Example: Tell the time in 3 ways
3:45
45 minutes after 3
15 minutes to 4

Directions: Choose or write the correct answers to the questions.

1. Match the clocks

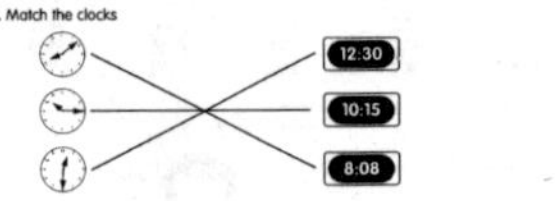

2. Tell the time in 3 ways.
   1:50 — 50 minutes after 1 — 10 minutes to 2
   11:33 — 33 minutes after 11 — 27 minutes to 12
3. Look at the clock. Show a half-hour later.
4. Look at the clock. Show 15 minutes earlier.

page 51

Skills: reads and interprets temperature in both Centigrade and Fahrenheit.

Directions: Write or choose the correct answers to the questions.

This is the thermometer John watched today.

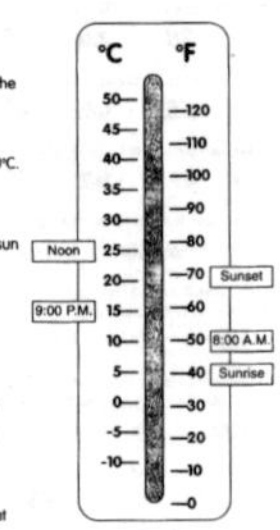

1. When John got up at 8:00 A.M. the thermometer read 50°F; What was that on the Centigrade scale? 10 °C
2. The highest temperature of the day was 30°C. Was it hotter than 90°F? NO
3. What was the temperature in °F when the sun went down? 70 °F
4. How much cooler (°C) was it at 9:00 P.M than at noon? 10 °C
5. How much higher was the temperature at sunset than at sunrise (°F)? 30 °F
6. John went to bed at 9:00 P.M. Was it warmer or colder than when he woke up at 8:00 A.M.? colder
   How many degrees Centigrade? 5 °C

page 52

Skills:
- counts money in coins and bills to $10
- makes change for $1.00

Directions: Write or choose the correct answer for each question. Study the pictures. Then answer the questions.

1. How much money does each person have?
   Saul has $ 2.77 Perry has $ 2.62
2. Saul buys gum for 75 cents. If he pays with a $1.00 bill, how much change will he get?
   a. 75 cents b. $1.00 (c. 25 cents) d. 50 cents
3. Perry bought a banana and an apple for a total of 83 cents. How much change should he get for a $1.00 bill?
   a. 27 cents (b. 17 cents) c. 33 cents d. 83 cents

Draw the coins Perry could get in change.

page 53

Skills:
- identifies and compares fractional parts through 10ths
- identifies fractions equal to 1

Directions: Choose or write the correct answers to the questions.

1. What fraction is shaded?
   2/3 5/8 3/4
2. Put the fractions in order from largest to smallest.
   $\frac{1}{4}$ $\frac{1}{5}$ $\frac{1}{9}$ $\frac{1}{2}$ $\frac{1}{3}$ $\frac{1}{10}$
   Smallest 1/10 1/9 1/5 1/4 1/3 1/2 Largest
3. Circle the fractions that are equal to 1:
   ($\frac{2}{2}$) $\frac{7}{8}$ $\frac{1}{3}$ $\frac{9}{10}$ ($\frac{4}{4}$) ($\frac{7}{7}$)
4. Use objects to help you solve these problems:
   a. $\frac{1}{4}$ of 16 is 4 b. $\frac{1}{2}$ of 10 is 5
   c. $\frac{1}{3}$ of 12 is 4 d. $\frac{1}{5}$ of 10 is 2

page 54

Skill: interprets information from picture and bar graphs

Directions: Write or choose the correct answers to the questions. Study the graphs. Then answer the questions.

Favorite Fruits

1. Which fruit did 7 students prefer? orange
2. Which fruit is most popular? banana
3. How many more students like bananas than like apples? 7 more

Favorite Pets

1. What pet do half the people own? dog
2. What is the least favorite pet? bird

page 55

Skills: knows multiplication and division facts through the nines

Directions: Do not use your calculator for this page.

1. Multiply
   $6 \times 6 = 36$ $9 \times 6 = 54$ $8 \times 7 = 56$
   $7 \times 7 = 49$ $8 \times 8 = 64$ $8 \times 9 = 72$
2. Divide
   $54 \div 6 = 9$ $48 \div 8 = 6$ $72 \div 9 = 8$
   $81 \div 9 = 9$ $63 \div 7 = 9$ $36 \div 6 = 6$
3. Multiply
   $7 \times 6 = 42$ $9 \times 7 = 63$ $9 \times 9 = 81$ $8 \times 6 = 48$
4. Divide
   $9\overline{)63}$ 7 $9\overline{)54}$ 6 $6\overline{)48}$ 8 $8\overline{)72}$ 9

page 56

Skill: uses multiplication and division facts through the 9's to solve word problems

Directions: Write and solve each problem in the space provided. Do not use a calculator. Write your answer on the line. Be sure to label your answer. For example:

Bob, Carlos, and Frank want to evenly divide 15 marbles. How many marbles will each get?
$15 \div 3 = 5$ 5 marbles each

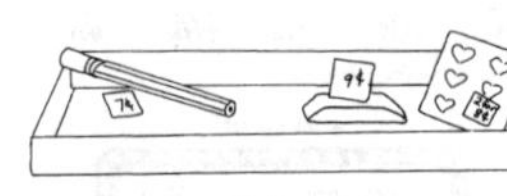

1. How much to buy 6 pencils? 42¢
2. How much to buy 12 stickers? 48¢
3. I have 55¢. Which of these can I buy - 7 erasers or 7 pencils? 7 pencils
4. Angela has 64¢. How many stickers can she get? 16 stickers
5. I want to buy 3 pencils and 2 erasers. How much money do I need? 39¢

These are the coins I have. Do I have enough money? Yes

page 57

Skill: identifies information needed and not needed to solve a problem

Directions: Read each story carefully. Then answer the questions.

1. On Saturday, Lavelle went to the beach with her friend, Karina. Lavelle is 9 years old and Karina is 10. The girls left home at about 1:00 P.M. They stayed at the beach for 3 hours. At what time did they leave the beach?

   Which sentence is not needed to answer the question?
   (a.) Lavelle is 9 years old and Karina is 10.
   b. The girls left home at about 1:00 P.M.
   c. They stayed at the beach for 3 hours.

2. The factory makes cars every day. The colors of the cars are black, white, red and green. Black cars are the favorite of most Americans. How many days would it take the factory to make 500 cars?

   Which sentence needs to be added to answer the question?
   a. More men than women drive cars.
   b. White cars are easy to see at night.
   (c.) The factory can make 100 cars every day.

3. The school choir will have a concert in the auditorium. The auditorium holds 300 chairs. How many rows of chairs will there be for the concert?

   Which sentence needs to be added to answer the question?
   a. 275 people will come to the concert.
   (b.) There will be 50 chairs in each row.
   c. 125 children will sing.

page 58

Skills:
- multiplies a two digit number by a one digit number with no regrouping (carrying)
- solves one step problems involving multiplication

Example: $30 \times 3 = 90$

Directions: Solve the problems below. Do not use a calculator.

1. $42 \times 2 = 84$ $14 \times 2 = 28$ $33 \times 3 = 99$ $20 \times 4 = 80$
2. $11 \times 7 = 77$ $10 \times 6 = 60$ $23 \times 2 = 46$ $32 \times 3 = 96$
3. There are 6 people in the White family. Each of them ate 11 pretzels as a snack. How many pretzels were eaten?
   66 pretzels
4. Each of the 24 kids in my class gets 2 new pencils every month. How many pencils are used each month?
   48 pencils
5. When Robert goes to the playground, he walks 4 blocks round trip. How many blocks will he walk if he goes to the playground 12 times?
   48 blocks

page 59

Skill:
- multiplies a two digit number by a one digit number with regrouping (carrying)
- solves one step problems involving multiplication

Example: $34 \times 4 = 136$

Directions: Work each problem or choose the right answer.

1. $67 \times 5 = 335$ $43 \times 8 = 344$ $19 \times 7 = 133$ $36 \times 3 = 108$ $45 \times 5 = 225$
2. $28 \times 7 = 196$ $14 \times 5 = 70$ $29 \times 3 = 87$ $17 \times 7 = 119$ $34 \times 3 = 102$
3. Meadowview School has 5 classrooms at each grade. Each classroom has 28 students. How many first graders will go to Meadowview School?
   a. 33 b. 28 c. 100 (d. 140)
4. Four school buses bring children to Meadowview School. Each bus can carry 56 children. How many children can ride the bus to school?
   a. 114 (b. 224) c. 60 d. 204

page 60

Parents: The skills on this page are challenging and may not have been taught in your child's third grade class.

Skills:
- divides a two-digit number by a one-digit number with and without remainders
- solve two-step word problems

Directions: Solve the problems below. Do not use a calculator

1. $8\overline{)30}$ 3 R6 $7\overline{)44}$ 6 R2 $5\overline{)26}$ 5 R1 $9\overline{)59}$ 6 R5
2. $4\overline{)56}$ 14 $6\overline{)90}$ 15 $3\overline{)72}$ 24 $5\overline{)65}$ 13

Directions: Solve these word problems. You may use a calculator to help you.

1. You have decided to buy a new video game that is on sale for $39.95. Your bank contains $17.75. If you can earn $4.50 a week walking the neighbors dog, how many weeks will it be before you can buy the video?

   work space:

   5 weeks

2. Super Save Food Market is selling ice cream for $2.49 a quart. Best Buy Groceries is selling ice cream for $1.69 a pint. Where will you go for the best buy? How much will you save if you buy a half gallon?

   work space:

   I will go to Super Save.
   I will save $1.78 on a half gallon.